ENERGY SCIENCE, ENGINEERING AND TECHNOLOGY

SOLAR POWER

TECHNOLOGIES, ENVIRONMENTAL IMPACTS AND FUTURE PROSPECTS

Energy Science, Engineering and Technology

Additional books in this series can be found on Nova's website under the Series tab.

Additional e-books in this series can be found on Nova's website under the e-book tab.

ENERGY SCIENCE, ENGINEERING AND TECHNOLOGY

SOLAR POWER

TECHNOLOGIES, ENVIRONMENTAL IMPACTS AND FUTURE PROSPECTS

STEPHEN BAILEY
EDITOR

nova publishers
New York

For permission to use material from this book please contact us:
Telephone 631-231-7269; Fax 631-231-8175
Web Site: http://www.novapublishers.com

Library of Congress Cataloging-in-Publication Data

Solar power : technologies, environmental impacts and future prospects / editors, Stephen Bailey.
pages cm -- (Energy science, engineering, and technology)
Includes bibliographical references and index.
ISBN 978-1-63321-317-3 (hardcover)
1. Solar energy. I. Bailey, Stephen (Stephen R.)
TJ810.S7174 2014
621.31'244--dc23

2014021850

Published by Nova Science Publishers, Inc. † New York

CONTENTS

PREFACE

Solar and other renewable power can play an important role in future electricity system, as shown by numerous energy and electricity system model studies. Many of these studies have in common that they focus on cost as the central issue, either by optimizing it or using it as a key variable in simulation. This book examines the renewable energy policies in Germany, the Netherlands, Spain, Switzerland and the UK in order to identify the policy motivations and constraints for expanding renewables in these countries. It also discusses fabrication, electronic properties and microstructure of TiO2-based dye-sensitized solar cells; solar tower power plants performance and reliability analysis; liquid crystal devices for photovoltaic systems; and mechanisms for solar technologies.

Chapter 1 – The increase in the demand for energy, water, materials and agricultural products have raised the concern about the emissions and waste flows associated with their production and use, and the future availability of resources.

In this chapter I discuss the material energy nexus by analysing the impacts of the increasing market share of Photovoltaic (PV) solar technologies on the future demand and supply of the metals required for these technologies, and the associated energy demand and environmental impacts. The analysis is carried out on a global level using a dynamic material flow and stock model and six scenarios for the demand for electricity and the electricity mix. Three scenarios are developed by International Energy Agency and three developed by European Renewable Energy Council and Greenpeace.

Several factors that have direct impact on the material demand are analysed, including the efficiencies of technologies, installation's geographical location, metals utilization and recycling rates, and the possibility of substitution.

In terms of resources and production capacity, the limiting metal for PV solar technologies is tellurium and to a lesser degree indium. Although secondary resources are expected to cover part of the demand, it is necessary to find other sources for the metals supply in the short to medium terms. Increasing technologies efficiencies, metals utilization rate, and choosing a good location would minimize the amount of metals required for PV solar technologies, however, some scenarios require more than double the available metals reserves. For these scenarios, it is necessary to either find new PV solar technologies that use different metals or find new sources for the metals used in current available technologies. The energy required to produce the metals and to treat the waste flows, and the associated impacts require carful considerations.

Chapter 2 – Solar and other renewable power can play an important role in the future European electricity system, as shown by numerous energy and electricity system model studies. Many of these reports have in common that they focus on cost as the central issue, either by optimizing it or using it as a key variable in simulation. In this paper, author examine the renewable energy policies in Germany, the Netherlands, Spain, Switzerland and the UK in order to identify the policy motivations and constraints for expanding renewables in these countries, in order to find out whether this cost-focus reflects actual policy priorities. Author find that their renewable energy policies are motivated by climate concerns and international agreements, energy security concerns, as well as job creation, and other macroeconomic or industry policy reasons. They are constrained primarily by short-term costs of renewable electricity support and grid integration challenges. Thus, actual policies are not very concerned with overall system costs and yet this is the issue many researchers focus on in their scenario building. Author contend that the political and scientific debates, including scientific system modelling, on deployment of renewable electricity should go beyond mere cost optimisation studies and focus instead on issues that actually drive and constrain policies today.

Chapter 3 – TiO_2-based dye-sensitized solar cells were fabricated and characterized Solar cells mixed with xylenol orange and rose Bengal showed higher conversion efficiency compared to solar cells with a single dye. In addition, amorphous TiO_2 layers were introduced, which improved the conversion efficiency. Microstructures of the solar cells were examined by X-ray diffraction and transmission electron microscopy, and energy levels of the present solar cells were also discussed.

Chapter 4 – The objective of the present study is to address the main methods and tools used to analyze the reliability and the performance of solar tower power plants (STPP).

An overview of STPP technology, current status and a presentation of the major components including the heliostat field, the solar receiver and the power conversion system are first highlighted. Next, the analysis of the performances and the reliability of STPP are outlined. Artificial neural network method (ANN) is used to assess the perfor-mance of STPP. In this study, the STPP receiver efficiency is considered as a performan-ce metric. In order to show the efficacy of this method, an example of hypothetical STPP is proposed. After the investigation of the simulation result, a very good learning and accuracy of ANN are noted. The reliability analysis methods are also investigated. The different models and tools, such as sofwares, for STPP reliability and performance evaluation are presented. The study will focus on specific methods such as the First Order Reliability Method (FORM) for the reliability and performance analysis of STPP. Detailed analysis concerning FORM is developed in this chapter. The hypothetical STPP example is also proposed to evaluate this method. The results show that FORM is suitable to assess the reliability of STPP systems and it converges quickly. It can also be used to adjust the first design parameters to a new safe design.

Chapter 5 – The power conversion efficiency, which is defined as the ratio of the output electrical power density to the incident light power density of the solar cell, is one of the most important parameters of a photovoltaic (PV) system. The efficiency depends on the photon flux density of the incident light, which changes frequently in different environment. For example, the solar cell in a solar power station receives the sunlight, which changes with the weather, season, and time; the solar cell on a mobile device may be used indoor and outdoor alternately. In order to preserve a static output of a solar cell in different environments, a tunable light modulator for keeping a static incident photon flux density is needed. Liquid crystal (LC) light modulators have been developed for years, which can be used to improve the PV systems. In this chapter, author introduce the PV systems with high and steady output by adopting electrically tunable LC light modulators. This concept may open a window to enhance the efficiency of CPV systems by using active optical elements.

Chapter 6 – In this work, an optimisation model for the integration of renewable energy sources for power generation (RES-E) technologies in power generation systems on a unit commitment basis is developed. The purpose of the optimization procedure is to account for RES-E net-metering scheme as well as for RES-E capacity auctions. The optimization model developed uses a genetic algorithm (GA) technique for the calculation of the required RES-E levy (or green tax) in the electricity bills. Also, this GA procedure enables the estimation of the level of the adequate (or eligible) feed-in-tariff to be offered to future RES-E systems, which do not participate in the net-metering and/or capacity auctioning procedure. In order to demonstrate the applicability of the optimisation procedure developed the case of rooftop residential photovoltaic systems are investigated for the net-metering option and large commercial PV systems are examined in the case of capacity auctioning scheme. The results indicated that the required green tax, in order to promote the use of RES-E technologies, which is charged to the electricity customers through their electricity bills, is reduced with the reduction in the final auctioning price and the degree of net-metering penetration. This has a significant effect related to the reduction of electricity bills.

In: Solar Power
Editor: Stephen Bailey
ISBN: 978-1-63321-317-3

Chapter 1

MATERIAL ENERGY NEXUS: THE IMPACTS OF METALS SUPPLY AND DEMAND ON THE FUTURE SOLAR POWER TECHNOLOGIES

Ayman Elshkaki[*]
Center for Industrial Ecology, School of Forestry and Environmental Studies, Yale University, New Haven, CT, US

ABSTRACT

The increase in the demand for energy, water, materials and agricultural products have raised the concern about the emissions and waste flows associated with their production and use, and the future availability of resources.

In this chapter I discuss the material energy nexus by analysing the impacts of the increasing market share of Photovoltaic (PV) solar technologies on the future demand and supply of the metals required for these technologies, and the associated energy demand and environmental impacts. The analysis is carried out on a global level using a dynamic material flow and stock model and six scenarios for the demand for electricity and the electricity mix. Three scenarios are developed by International Energy Agency and three developed by European Renewable Energy Council and Greenpeace.

[*] Corresponding author: Ayman Elshkaki. Center for Industrial Ecology, School of Forestry and Environmental Studies, Yale University, New Haven, CT 06511, US. Tel.: +1-302-4364246, fax: +1-302-4325556, e-mail address: ayman.elshkaki@yale.edu.

Several factors that have direct impact on the material demand are analysed, including the efficiencies of technologies, installation's geographical location, metals utilization and recycling rates, and the possibility of substitution.

In terms of resources and production capacity, the limiting metal for PV solar technologies is tellurium and to a lesser degree indium. Although secondary resources are expected to cover part of the demand, it is necessary to find other sources for the metals supply in the short to medium terms. Increasing technologies efficiencies, metals utilization rate, and choosing a good location would minimize the amount of metals required for PV solar technologies, however, some scenarios require more than double the available metals reserves. For these scenarios, it is necessary to either find new PV solar technologies that use different metals or find new sources for the metals used in current available technologies. The energy required to produce the metals and to treat the waste flows, and the associated impacts require carful considerations.

1. Introduction

The increase in the global population and the economic growth has been associated with an increase in the demand for energy, water, materials and agricultural products. This has raised the concern about the emissions and waste flows associated with their production and use, and the future availability of resources.

The four systems, energy, water, agriculture, and materials, are coupled in a variety of ways. In recent years a number of studies have attempted to address the nexus of these systems (Ackerman and Fisher, 2013; Bazilian et al., 2011; Li et al., 2012; Siddiqi and Anadon, 2011). Most of these studies have focused on the energy, water and agriculture nexus, and a few on the materials required to sustain the development in these systems, mainly the metals required for the energy system (Elshkaki and Graedel, 2013; Moss et al., 2013; Kleijn et al., 2011).

Metals are essential for all energy generation technologies. They are either an important component of the technologies or used to enhance their efficiencies. Meanwhile, the mining industry is one of the most energy intensive sectors. Graedel (2011) defines the energy metals as "the metals for which the bulk of their use is energy related, or metals that are used in small amounts for energy applications but that are crucial to the success of those applications".

Among these are cadmium (Cd), silver (Ag), gallium (Ga), germanium (Ge), selenium (Se), indium (In), tellurium (Te), neodymium (Nd), and dysprosium (Dy).

In this chapter, I discuss the energy material nexus illustrated by the Photovoltaic (PV) solar technologies, the metals required for them, and the energy required for the production of these metals.

Solar power technologies are among the most promising technologies for the future electricity generation. Although solar power technologies have low carbon footprint compared to other energy production technologies (POST, 2011) and their learning rate is the highest in the energy world, there is a growing concern about the availability of the metals required for the future development and large scale implementation of these technologies and the environmental impacts associated with their production.

Recently, several studies have evaluated the metals required for PV solar technologies and identified a limiting metal for each technology: silver for silicon based PV technologies (mc-Si and sc-Si) and concentrated solar power (CSP), Germanium for amorphous silicon (a-Si) technology, Cadmium and Tellurium for cadmium telluride (CdTe), and Indium and Gallium for copper indium gallium diselenide (CIGS) (Andersson et al., 1998; Elshkaki a nd Graedel, 2013; Feltrin and Freundlich, 2008; Fthenakis, 2009; Wadia et al., 2009; Zuser and Rechberger, 2011; Jacobson and Delucchi, 2011; Kleijn and Van der Voet, 2010; Kleijn et al., 2011). Some of these metals have limited availability and identified as critical (Goe and Gaustad, 2014; US DOE, 2011) and all of them are mainly produced as companion metals from other metals deposits.

The magnitude of the metals required for PV technologies is mainly determined by the scenarios used for the future electricity demand and the market share of electricity generation technologies, and the assumptions made on the factors that affect the metal required for specific technology, including installation's geographical location, efficiency, performance ratio, metal's content and their utilization rates, possibility for substitution, the life time of the technology, and metals recycling rates.

Similarly, energy is required in all metals production stages (mining, mineral processing/Beneficiation, smelting, and refining). The global energy consumption for the main primary metals production is estimated to be 32 EJ/y in 2007 (about 18% of the world annual total electricity consumption) (Norgate and Jahanshahi, 2011). The energy required for the production of metals is mainly determined by the ore grade of the metals, energy efficiency, and demand for the metals.

In the following sections, the metals required for PV solar technologies and the energy required to produce them are discussed.

2. METHODOLOGY

2.1. Scenarios for the Future Electricity Generation and the Associated CO_2 Emissions

In recent years several scenarios have been developed for the future demand of primary energy, future demand of electricity, and the future market share of different electricity generation technologies (Electris, 2009; EREC and GP, 2010; EWEA, 2008; GWEC, 2010; IEA, 2011). These scenarios include Current Policy, New Policy, and 450 scenarios developed by the International Energy Agency (IEA, 2011), and Reference, Energy Revolution, and Advanced Energy Revolution scenarios developed by the European Renewable Energy council and Greenpeace (EREC and GP, 2010). The total electricity generation and the carbon dioxide emissions associated with power generation in different scenarios are shown in figure 1. The share of PV solar technologies in different scenarios ranges from 1.3% of the total electricity generation by 2050 in the Reference scenario to 16% in the Advanced Energy Revolution scenario. The electricity generated by PV solar technologies in different scenarios is shown in figures 2. Data on the three IEA scenarios are available until 2035. These values are extrapolated from 2035 to 2050.

2.2. Description of Photovoltaic (PV) Solar Technologies

Two types of PV technologies are commonly available in the market; crystalline silicon (including multi-crystalline (mc-Si) and single-crystalline (sc-Si)) and thin film (including amorphous silicon (a-Si), cadmium telluride (CdTe), and copper indium gallium diselenide (CIGS)).

The global installed capacity of PV solar has increased from 1.4 GW in 2000 to over 100 GW at the end of 2012 (EPIA, 2013).

The market share of thin film technologies in 2005 was 6.5% (Van Sark, 2007), increased to 10.4% in 2007 (EPIA and GP, 2008), and to 19% in 2009 (Zuser and Rechberger, 2011) with the remaining being crystalline silicon.

The life time of the PV technologies is between 20 and 30 years (Zuser and Rechberger, 2011).

a

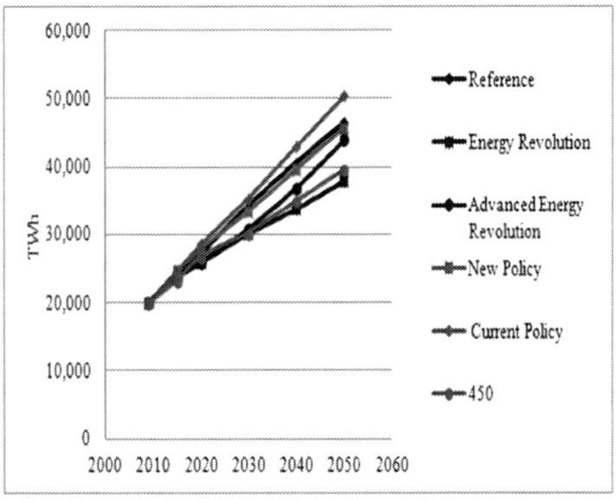

b

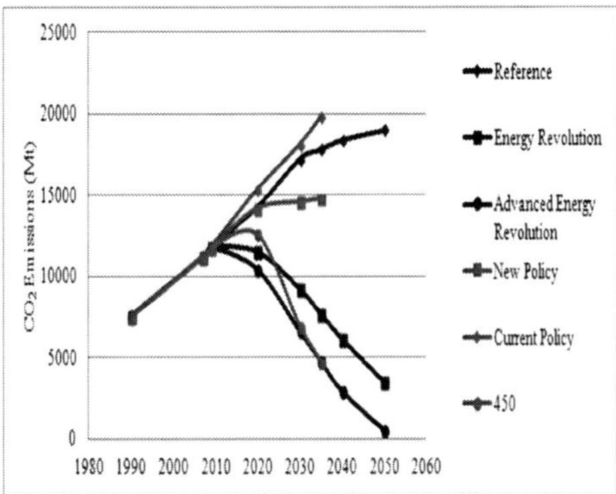

Figure 1. Total electricity generation (a) and carbon dioxide emissions associated with power generation (b) in different scenario.

In the model, a life time of 25 years is used. The average annual global irradiation is assumed 1500 KWh/m^2 based on solar irradiation values in 37 countries obtained from JRC (2012). The efficiencies of mc-Si, a-Si, CIGS, and CdTe in 2007 were 13%, 7%, 10%, and 10%, respectively (Raugei and Frankl, 2009). Those for CIGS and CdTe in 2008 were 11.2% and 10.8% (Fthenakis, 2009), and those for mc-Si and a-Si in 2010 were 14.6% and 9.3% (Zuser and Rechberger, 2011). The utilization rates of PV solar technologies are 50%, 40%, 40%, and 90% for the mc-Si, CIGS, CdTe, and a-Si PV solar technologies in 2010 (Zuser and Rechberger, 2011).

Losses due to conversion from direct current (DC) to alternating current (AC) and due to the transportation distance from the module to the grid are about 30%, expected to decrease to 20%.

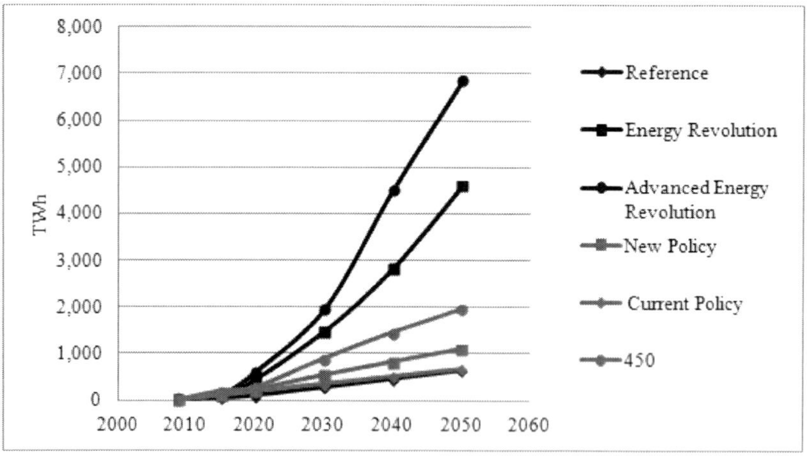

Figure 2. Electricity generated by PV solar technologies from 2010 to 2050 in different scenarios.

Table 1. The amount of metals used in PV solar power technologies

	Al	Cu	Ag	Pb	Fe	Ni	Cd	Te	Se	Ga	In	Ge
Si based PV (g/m^2)	2700	1110	52	3.10	20000	0.16						
CdTe (g/m^2)							6.30	6.50				
CIGS (g/m^2)									4.80	0.53	2.90	
aSiGe (g/m^2)												0.44

Andersson, 2000 and Ecoinvent, 2012.

In the model, the performance ratio is assumed to be 0.75 (JRC, 2012). The metals required for silicon-based and thin film technologies are listed in table 1 (Andersson and Jacobsson, 2000; Ecoinvent, 2012).

The future market share of Si based PV technologies is estimated based on data and information from different sources to account for 79% in 2012, 74% in 2015, 61% in 2020, and 50% in 2030 (Jager-Waldau, 2011 and EPIA and GP, 2011).

It is expected that by 2040 the efficiencies of Si based, a-Si, CIGS, and CdTe will reach 25%, 11.7%, 15.7%, and 16.7%, respectively (Zuser and Rechberger, 2011).

The utilization rates are expected to increase to 70%, 80%, and 80% for the mc-Si, CIGS, and CdTe PV solar technologies by 2040 (Zuser and Rechberger, 2011) and expected to remain 90% for a-Si thin film technology.

In the first scenario, these values are used in the model and extrapolated to 2050 with the share of Si based technologies reaching 12%, and the assumption that the mc-Si and sc-Si will have equal shares in the Si based PV market and the three thin film technologies will have equal shares in the PV thin film market.

By 2050, the efficiencies of Si based, a-Si, CIGS, and CdTe are expected to reach 30%, 12%, 19%, and 17.5% and the utilization rates are expected to reach 78%, 90%, 83%, and 83% respectively.

In the second scenario, it is assumed that the efficiencies of Si based, a-Si, CIGS, and CdTe technologies are expected to reach 35%, 16%, 23%, and 23% by 2050 respectively and the other factors to remain the same as in the first scenario. In the third scenario, it is assumed that the utilization rates of all PV technologies are expected to reach 90% and the other factors to remain the same as in the first scenario. In the forth scenario, it is assumed that the average annual global irradiation is 2000 $KWh/m^2$and the other factors remain the same as in the first scenario. The fifth scenario combines the values for the efficiencies, utilization rates and solar irradiation assumed in the second, third and fifth scenarios.

2.3. Dynamic Material Flow Analysis

The electricity generated by PV solar technologies from 1980 through 2050 is estimated based on the demand for electricity and the share of PV technologies in electricity generation.

The cumulative installed capacity (stock), annual installed capacity (inflow) and discarded flow of old installed capacity (outflow) are estimated as given by Equation 1 to Equation 5 (Elshkaki and Graedel, 2013).

The cumulative installed capacity of the PV solar technology is estimated based on the total electricity generated by the technology in each scenario.

The annual installed capacity for the technology is estimated based on the cumulative installed capacity and the annual discarded flow of old annual installed capacity as given by Equation 1. The discarded flow of PV solar technology is estimated as a delayed inflow as given by Equation 2 and set to zero from the first year of their installation to a time equal to their life time.

The annual installed capacity of each sub-technology (Si based, a-Si, CIGS, and CdTe) is estimated based on the annual installed capacity of the PV technology and the market share of each sub-technology as given by Equation 3.

The discarded flow of old annual installed capacity (outflow) of each sub-technology is estimated as a delayed annual installed capacity (inflow) of each sub-technology using an equation similar to Equation 2, and the cumulative installed capacity (stock) of each sub-technology is estimated as given by Equation 4.

The efficiency, the utilization rate, and the performance ratio are used after the estimation of the annual installed capacity of each sub-technology.

The inflow of metals in each sub-technology is estimated based on the inflow of the sub-technology multiplied by the metal content. The discard flow is estimated as a delayed inflow using an equation similar to Equation 2, and the stock is estimated based on an equation similar to Equation 4. If metal emissions occur during use, the flow is given by Equation 5. These flows are expected to be small compared to the discard flow, however.

$$F_T^{in}(t) = S_T(t+1) - S_T(t) + F_T^{out}(t) \tag{1}$$

in which F_T^{in} (t) is the annual installed capacity of the technology at time t (inflow), $S_T(t)$ is the cumulative installed capacity of the technology at time t (stock), and F_T^{out} (t) is the annual discarded flow of old installed capacity of the technology at time t (outflow).

$$F_T^{out}(t) = F_T^{in}(t-L) \tag{2}$$

in which L is the life time

$$F_{ST}^{in}(t) = \alpha * F_T^{in}(t) \tag{3}$$

$$S_{ST}(t+1) = S_{ST}(t) + F_{ST}^{in}(t) - F_{ST}^{out}(t) \tag{4}$$

where ST refers to sub-technology and α is the market share.

$$F_M^{out}(t) = F_M^{in}(t-L) * (1-\gamma)^L \tag{5}$$

where γ is the emission factor and M refers to metal.

2.4. Energy Required for Metals Production

The amount of energy required for metals production and consequently the emitted carbon dioxide are mainly determined by the ore grade of the metals, the efficiency of energy, the demand for the metals, and the energy mix in the producing countries. The energy required for metals production increases as a result of the decrease in ore grade (Mudd 2007 and 2010), while decreases as a result of the increase in energy efficiency. The actual energy currently used for the production of metals is more than the theoretical required energy (Alvarado et al., 1999; Norgate and Jahanshahi, 2011), which indicate the possibility for increasing the energy efficiency.

The total amount of energy required to produce the metals used in Si based PV solar technology is estimated based on the demand for these metals and the energy required for the production of each metal (Table 2).

3. Results and Discussions

3.1. Demand and Supply of Metals

The installed capacities of the four PV solar technologies in different scenarios are shown in figure 3. The annual demand for the metals required in these technologies is expected to increase over time except for Ag. The annual demand for Te is expected reach 377 Mg, 3070 Mg, 4693 Mg, 374 Mg, 667 Mg, and 1257 Mg by 2050 in the Reference, Energy Revolution, Advanced Energy Revolution, Current Policy, New Policy, and 450 scenarios.

Table 2. Total energy requirements for the production of metals from the ore (GJ/ton)

	Bodsworth, 1994	Rankin, 2011 and 2012	Ray et al., 2005
Al	220-250	211.5	300
Cu	95-110	39.3	85
Iron/steel	25-30	22.7	29
Pb	20-25	21.02	32
Ni	140-150	145.6	210 (from sulphide)
Zn	60-70	47.14	63
Ag		1710	

Bodsworth 1994; Rankin, 2011 and 2012; Ray et al., 2005.

For In, the demand is expected to reach 182 Mg, 1480 Mg, 2260 Mg, 180 Mg, 321 Mg, and 605 Mg by 2050. For Ge, the demand is expected to reach 37 Mg, 298 Mg, 455 Mg, 36 Mg, 65 Mg, and 121 Mg by 2050. The demand for Ag is expected decrease overtime reaching 873 Mg, 7106 Mg, 10860 Mg, 865 Mg, 1542 Mg, and 2910 Mg by 2050.

The average yearly demand for Te from 2014 to 2050 is expected to exceed its 2010 production in all scenarios and expected to be 1660% and 493% of its 2010 production (Table 3) in the Advanced Energy Revolution and 450 scenarios respectively. The average yearly demand for In is expected to exceed its 2010 production in the Energy Revolution and Advanced Energy Revolution scenarios only and expected to be 170% and 50% of its 2010 production (Table 3) in the Advanced Energy Revolution and 450 scenarios respectively. The average yearly demand for Ge is expected to exceed its 2010 production in the Energy Revolution and Advanced Energy Revolution scenarios only and expected to be 170% and 50% of its 2010 production (Table 3) in the Advanced Energy Revolution and 450 scenarios respectively. The average yearly demand for Ag is expected to exceed its 2010 production in the Advanced Energy Revolution scenario only and expected to be 104% and 33% of its 2010 production (Table 3) in the Advanced Energy Revolution and 450 scenarios respectively.

The cumulative demand for the metals up to 2050 is compared to the 2010 available reserves given in Table 3. The cumulative demand for Te, In, Ag, and Se from 2014 to 2050 compared to the 2010 reserve estimates in different scenarios is shown in figure 4, demonstrating the potential supply challenges for Te, In, and Ag in Energy Revolution, Advanced Energy Revolution, and 450 scenarios.

Table 3. Production, reserve and reserve base of the metals used in PV solar technologies

	Production (Mg)	Reserve (Mg)	Reserve Base (Mg)
Ag	22,200	510,000	570,000
Cd	22,000	660,000	
Ga	106	1,000,000	
Ge	120	450,000	500,000
In	574	11,000	16,000
Se	2,260	88,000	172,000
Te	123	22,000	48,000

US Geological Survey, 2009.

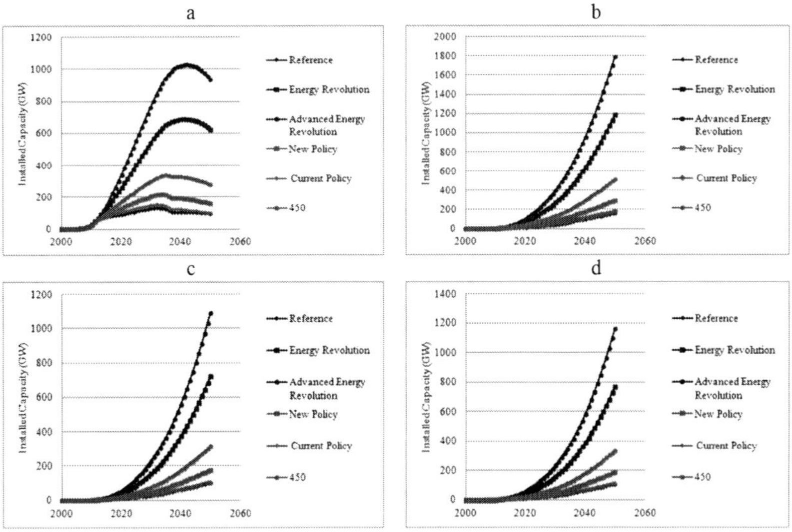

Figure 3. Installed capacities of Si based (a), a-Si (b), CdTe (c), and CIGS (d) from 2000 to 2050 in different scenarios.

In term of resources, the cumulative Ag demand in the next 37 years is either close to or slightly exceeding the total available reserves in the Energy Revolution and Advanced Energy Revolution scenarios. The Cumulative Ag demand is expected to be met by the available reserves in all other scenarios. The cumulative In and Te demand in the next 37 years is either close to or exceeding their available reserves in the Energy Revolution, Advanced Energy Revolution, and 450 scenarios. The Cumulative demand of the two metals is expected to be met by their available reserves in all other scenarios.

For Ge, Ga, Cd and Se, resource availability is not a problem. In all scenarios, the cumulative demand for Ge range between 0.2% and 1.7% of the available reserves, between 0.1% and 0.4% for Ga, between 1% and 11% for Cd, and between 6% and 67% for Se.

The demand for Te, In, and Ag by 2050 and their cumulative demand in different energy scenarios, and scenarios 1 to 5 are shown in figure 5. The cumulative demand for the three metals from 2014 to 2050 compared to the 2010 reserves estimates in different energy scenarios, and scenarios 1 to 5 are listed in table 4.

The amounts of Te and In required for CdTe and CIGS technologies are expected to decrease from 123 kg/MW in 2010 to 60 kg/MW in 2050 and from 55 kg/MW to 26 kg/MW in the first scenario.

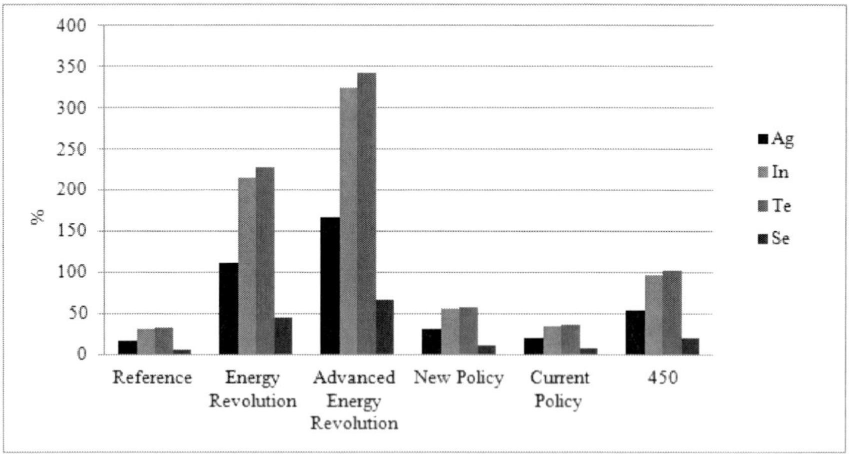

Figure 4. Cumulative demand for Te, In, Ag, and Se from 2014 to 2050 compared to the 2010 reserves in different scenarios.

The amounts of the two metals are expected to reach 60 kg/MW and 26 kg/MW by 2050 in the second scenario. These amounts are expected to reach 55 kg/MW and 24 kg/MW in the third scenario, 46 kg/MW and 20 kg/MW in the forth scenario, and 42 kg/MW and 19 kg/MW in the fifth scenario. In the second, third, and forth scenario, the cumulative Ag demand in the next 37 years is expected to exceed the total available reserves in the Advanced Energy Revolution scenario only, while in the fifth scenario, the cumulative Ag demand can be met by the available reserves in all scenarios. In scenarios two to five, In and Te cumulative demand is expected to exceed the total available reserves in the Energy Revolution and Advanced Energy Revolution scenarios only, however not as much as in scenario 1.

3.2. The Impact of Recycling

To assess the impact of recycling, a recycling rate of 90% is assumed for all metals, and the values in scenario 5 are used for the efficiency, utilization rate, and solar irradiation. Secondary resources is expected to cover about 14% of In and Te average annual demand in the Energy Revolution and Advanced Energy Revolution Scenarios and between 18% and 25% in the three IEA scenarios. The cumulative demand of In compared to 2010 reserve is expected to be reduced from 125% and 187% to 107% and 163% in the Energy Revolution and Advanced Energy Revolution Scenarios respectively.

Table 4. Cumulative demand for Ag, In, and Te from 2014 to 2050 compared to the 2010 reserves (%) in different scenarios

	Reference			Energy Revolution			Advanced Energy Revolution			Current Policy			Current Policy			450		
	Ag	In	Te	Ag	In	Te	Ag	In	Te	Ag	In	Te	Ag	In	Te	Ag	In	Te
SC1	16	30	32	**111**	**215**	**228**	**167**	**325**	**343**	20	33	35	31	55	58	53	96	**102**
SC2	14	25	28	**100**	**175**	**197**	**150**	**264**	**296**	18	27	31	28	45	50	48	79	88
SC3	14	28	30	98	**198**	**210**	**146**	**297**	**315**	17	30	33	28	50	54	47	89	94
SC4	12	23	24	86	**166**	**176**	**129**	**250**	**265**	15	26	27	24	42	45	41	74	79
SC5	10	18	20	68	**125**	**140**	**101**	**187**	**210**	12	20	22	19	32	36	33	56	63

For Te, the cumulative demand compared to 2010 reserve is expected to be reduced from 140% and 210% to 120% and 182% in the two scenarios respectively.

3.3. Substitution

Metals are used in PV solar technologies for their physical and chemical properties and are not substituted. Although there are no satisfactory substitutes on the material level, the different PV technologies could be seen as possible substitutes to each other.

In recent years, the PV solar technologies are growing rapidly but there is no clear dominate technology. Therefore in the analysis presented in sections 3.1 and 3.2, it is assumed that three thin film technologies will have equal shares in the PV thin film market.

To assess the impact of substitution, we assume 50% market share of CIGS in the thin film market, 17% for CdTe, and 33% for a-Si. These values constitute 44%, 15%, and 29% of the PV market for CIGS, CdTe, and a-Si respectively. The values in scenario 5 are used for the efficiency, utilization rate, and solar irradiation.

The cumulative demand of In compared to 2010 reserves is expected to increase from 125% and 187% to 186% and 280% in the Energy Revolution and Advanced Energy Revolution Scenarios respectively.

For Te, the cumulative demand compared to 2010 reserves is expected to decrease from 140% and 210% to 70% and 105% in the two scenarios respectively. The results show that Te cumulative demand based on this market share of CdTe can be met by the available reserves in all scenarios.

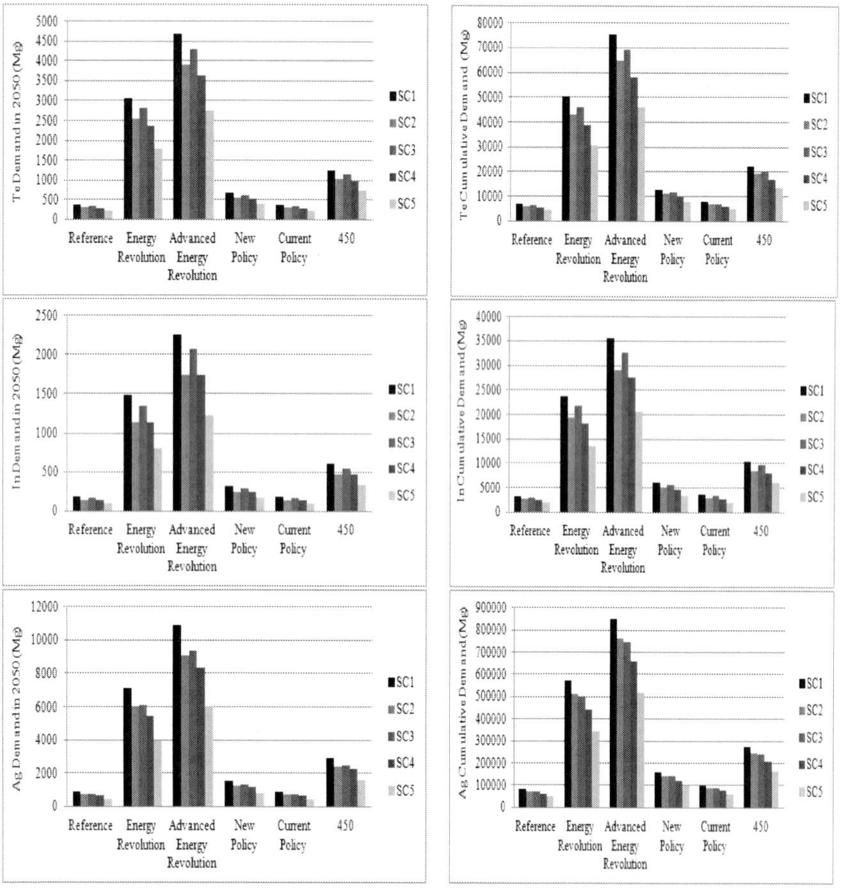

Figure 5. The demand for Te, In, and Ag by 2050 and their cumulative demand from 2014 to 2050 in different scenarios.

The results also suggest that substituting one technology by another would reduce the pressure on one metal and increase it on the other one.

3.4. Coproduction Issues

Co-production is the production of two or more metals from the same ore deposit. The metals with high concentration are refered to as the "hosts" and those with low concentrations as "companions".

Most of the metals that are essential for PV solar technologies are companion metals.

In, Ge, and Cd are mainly produced from Zn deposit, and Te and Se from Cu deposit. Typical Zn and In concentrations in the Zn deposit are 7.5% and 0.001% (BGR, 2012) and typical ore grades of Cu and Te in copper deposits are 0.8% and 0.0002% (Fthenakis and Anctil, 2013). To increase the production of companion metals, it is necessary either to increase their recycling, increase their recovery rate from ores, or increase the production of the host metals. It has been show in the analysis above that secondary materials could cover only small fraction of the demand for the two metals and the majority of the demand especially in the next 25 years has to come from primary resources due to the life time of the technologies. In 2010, the global ratios of In to Zn and Te to Cu production were 50g of In for each ton of Zn and 6.7 g of Te for each ton of Cu (USGS, 2011). The possible theoretical ratios based on the concentrations in the deposit are 133g of In per ton of Zn and 250 g of Te for each ton of Cu. This is only 38% of In available in the Zn deposit and 2.7% of Te available in the Cu deposit. Although increasing the recovery rate of companion metals is an option, it requires building a new production facility and an increase in the price of the metals. Increasing the production of the host metals is the easiest but would have consequences on the supply of host metals from secondary resources (Elshkaki and Van der Voet, 2006). If the demand for In or Te is met by the increase of Zn and Cu production, the supply of the host metals from primary resources might exceed their demand leading to an oversupply of these metals. This will have direct impact on the energy required to produce these metals as the energy required to produce them from secondary resources requires less energy (Rankin, 2011). Another important issue related to the coproduction of metals is the production of Cd. Cd is also companion metal that is mainly produce from Zn deposits. The impact of the increase in the In demand for CIGS is not limited to the oversupply of Zn but also to a possible oversupply of Cd. This amount needs to be carefully disposed with an additional energy requirement. On the other hand, the use of Cd in CdTe solar technology would partly resolve the problem associated with the oversupply of Cd.

3.5. Energy Required for Metals Production

The analysis of the energy required for the metals used in Si based is based on a scenario that assumes 75% market share for the technology from 2015 onward, 75% progress ratio, 1540 KWh/m^2 average annual global irradiation, 30% efficiency by 2050, and 78% utilization rate by 2050.

Based on these values, the cumulative demand for Ag is expected to be less than the available reserves in the Current Policy, New Policy, and Reference scenarios. The cumulative demand for Ag is expected to be 100%, 220%, and 325% of the available reserves in the 450, Energy Revolution, and Advanced Energy Revolution scenarios respectively.

Based on the demand for metals and the energy required for the production of each metal (Table 4), the total amount of energy required for all metals by 2050 is expected to reach 123 PJ, 1010 PJ, 1540 PJ, 123 PJ, 220 PJ, and 415 PJ in the Reference, Energy Revolution, Advanced Energy Revolution, Current Policy, New Policy, and 450 scenarios respectively.

These amounts are about 0.03%, 0.26%, 0.34%, 0.03%, 0.05, and 0.1% of the expected electricity generation in 2050. The cumulative amount of energy required for the production of all metals in the next 37 years in different scenarios is shown in figure 6.

The emissions of CO_2 will be determined by the total energy required and the sources of electricity supply. For the metals that require a high component of electrical energy, the change in the market share of electricity generation technologies, such as those in the Energy Revolution, Advanced Energy Revolution, and 450 scenarios, will have huge impact on the CO_2 emissions.

CONCLUSION

This chapter discusses the material energy nexus by analysing the impacts of the increasing market share of renewable energy technologies on the future demand and supply of the metals required for these technologies, and the associated energy demand and environmental impacts.

The analysis is carried out for the materials required for the PV solar power technologies (Si based, a-Si, CdTe, CIGS) on a global level using a dynamic material flow and stock model and six scenarios for the demand for electricity and the electricity mix.

Three of these scenarios are developed by International Energy Agency (IEA) and three scenarios developed by European Renewable Energy Council and Greenpeace (EREC and GP).

Several factors that have direct impact on the material demand are analysed, including the efficiencies of the technologies, the installation geographical location, the metals utilization rate, recycling rate and the possibility of substitution.

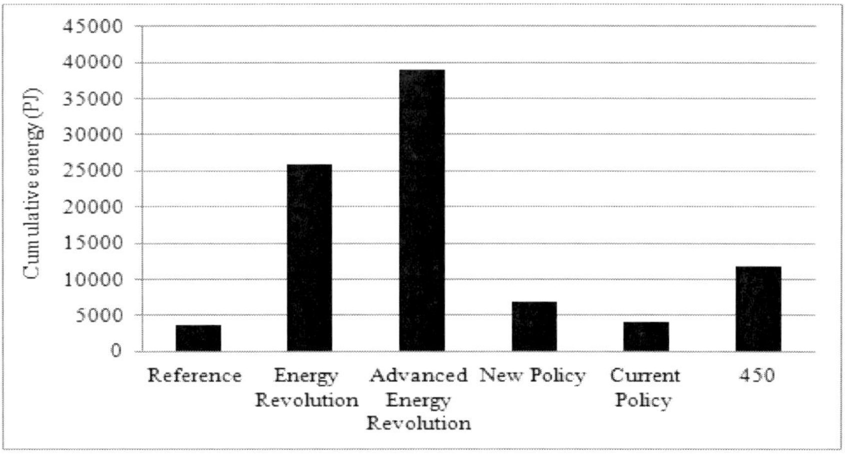

Figure 6. Cumulative energy required for the production of all metals from 2014 to 2050 in different scenarios.

Several conclusions can be drawn from the analysis.

- No resource problems occur in the availability of metals required for PV solar technologies in the IEA scenarios. In contrast, two of the three EREC and GP scenarios have resource problems.
- In terms of resources and production capacity, the limiting metal for PV solar technologies is tellurium and to a lesser degree indium.
- The future electricity generation technologies will be largely determined by the availability of materials. To resolve the problems associated with limited availability of metals, it is necessary to either increase their recovery rate from ores or carefully adjust the market share of renewable technologies in the future.
- Although increasing technologies efficiencies, metals utilization rate, and choosing a good location would minimize the amount of metals required for PV solar technologies, some scenarios require more than double the available metals reserves. For these scenarios, it is necessary to either find new PV solar technologies that use different metals or find new sources for the metals used in current available technologies.
- Although the recycling rates for the metals required for PV solar technologies are expected to be high, secondary resources are not expected to cover a big fraction of the total demand in the short to medium terms.

- The substitution of one PV technology by another would move the pressure from one metal to another.
- Although solar power technologies have low carbon footprint compared to other electricity production technologies, the energy required to produce the metals and to treat the waste flows, and the associated impacts require carful considerations.
- The analysis carried in this chapter does take into account other competing technologies either in other electricity generation or in other sectors. A complete analysis including these competing technologies is required.

REFERENCES

Ackerman, F., Fisher, J., 2013. Is there a water–energy nexus in electricity generation? Long-term scenarios for the western United States. *Energy Policy*, 59, 235-241.

Andersson, B. A., Azar, C., Holmberg, J., Karlsson, S., 1998. Material constraints for Thin-Film solar cells. *Energy*, 23, 407–411.

Andersson, B. A., Jacobsson, S., 2000. Monitoring and assessing technology choice: the case of solar cells. *Energy Policy*, 28, 1037–1049.

Alvarado, S., Maldonado, P., Jaques, I., 1999. Energy and environmental implications of copper production, *Energy*, 24, 307-316.

Bazilian, M., Rogner, H., Howells, M., Hermann, S., Arent, D., Gielen, D., Steduto, P., Mueller, A., Komor, P., Tol, R. S. J., Yumkella, K. K. 2011. Considering the energy, water and food nexus: Towards an integrated modeling approach. *Energy Policy*, 39, 7896-7906.

Bodsworth, C., 1994. *The extraction and refining of metals*. CRC Press Inc., Florida, US.

Ecoinvent Database v2.2, 2012. *Swiss Centre for Life Cycle Inventories*, Duebendorf, Switzerland. Available at: http://www.ecoinvent.ch.

Electris, C., Raskin, P., Rosen, R., Stutz, J., 2009. *The century ahead: Four global scenarios. Technical documentation*. Tellus Institute. Boston, US.

Elshkaki, A., Graedel, T. E., 2013. Dynamic analysis of the global metals flows and stocks in electricity generation technologies, *Journal of Cleaner Production*, 59, 260-273.

Elshkaki, A., Van der Voet, E., 2006. The consequences of the use of platinum in new technologies on its availability and on other metals cycle. In:

Loeffe, C. V., Editor. *Conservation and Recycling of Resources: New Research,* New York: Nova Science Publishers, Inc; p. 61-92.

European Photovoltaic Industry Association (EPIA), 2013. *Global market outlook for photovoltaic 2013-2017.* Brussels, Belgium.

EPIA and GP, 2008. Solar Generation V – 2008: Solar Electricity for over One Billion People and Two Million Jobs by 2020. European Photovoltaic Industry Association and Greenpeace.

EPIA and GP, 2011. Solar Generation 6 – 2011: Solar Photovoltaic Electricity Empowering the World. European Photovoltaic Industry Association and Greenpeace.

European Renewable Energy Council and Greenpeace (EREC and GP), 2010. *Energy Revolution: A sustainable Global Energy Outlook.*

EWEA, 2008. *Pure power – Wind energy scenarios up to 2030, Final report.* European Wind Energy Association, Brussels, March.

Federal Institute for Geosciences and Natural Resources (BGR), *2012. Fact Sheet: Indium, POLINARES working paper no. 39.* Available at http:// www.polinares.edu.

Feltrin, A., Freundlich, A., 2008. Material considerations for terawatt level deployment of photovoltaics. *Renewable Energy*, 33, 180–185.

Fthenakis, V., 2009. Sustainability of photovoltaics: the case for thin-film solar cells. *Renewable and Sustainable Energy Reviews*, 13, 2746–2750.

Fthenakis, V., Anctil, A., 2013. Direct Te mining: Resource availability and impact on cumulative energy demand of CdTe PV life cycles. *IEEE Journal of Photovoltaics*; 3, 433.

Goe, M., Gaustad, G., 2014. Identifying critical materials for photovoltaics in the US: A multi-metric approach. *Applied Energy*, DOI. http://dx.doi.org/ 10.1016/j.apenergy.2014.01.025.

Graedel, T. E., 2011. On the availability of the energy metals. *Annul Review Materials Research*, 41, 323–335.

GWEC, 2010. *Global wind energy outlook 2010.* Global Wind Energy Council and Greenpeace International.

International Energy Agency, 2011. *World Energy Outlook 2011*. Available at: www.iea.org.

Jacobson, M. Z., Delucchi, M. A., 2011. Providing all global energy with wind, water, and solar power, Part I: technologies, energy resources, quantities and areas of infrastructure, and materials. *Energy Policy*, 39, 1154–1169.

Jager-Waldau, A., 2011. *PV Status Report: Research, Solar Cell Production and Market Implementation of Photovoltaics*. European Commission, DG

Joint Research Centre, Institute for Energy, Renewable Energy Unit, Ispra, Italy.

JRC, 2012. Available at: http://re.jrc.ec.europa.eu/pvgis/index.htm.

Kleijn, R., Van der Voet, E., 2010. Resource constraints in a hydrogen economy based on renewable energy sources: An exploration. *Renewable and Sustainable Energy Reviews*, 14, 2784–2795.

Kleijn, R., Van der Voet, E., Kramer, G. J., Van Oers, L., Van der Giesen, C., 2011. Metal requirements of low-carbon power generation. *Energy*, 36, 5640–5648.

Li, X., Feng, K., Siu, Y. L. Hubacek, K., 2012. Energy-water nexus of wind power in China: The balancing act between CO_2 emissions and water consumption. *Energy Policy*, 45, 440-448.

Moss, R. L., Tzimas, E., Willis, P., Arendorf, J., Espinoza, L. T. 2013. Critical Metals in the Path towards the Decarbonisation of the EU Energy Sector: Assessing Rare Metals as Supply-Chain Bottlenecks in Low-Carbon Energy Technologies. *Report EUR 25994 EN, European Commission,* Publications Office of the European Union, Luxembourg.

Mudd, G. M., 2007. An analysis of historic production trends in Australian base metal mining, *Ore Geology Reviews*, 32, 227–261.

Mudd, G. M., 2010. The environmental sustainability of mining in Australia: key mega-trends and looming constraints, *Resources Policy*, 35, 98–115.

Norgate, T. and Jahanshahi, S. 2011. Reducing the greenhouse gas footprint of primary metal production: Where should the focus be? *Minerals Engineering,* 24, 1563-1570.

POST, 2011. Carbon Footprint of Electricity Generation. Houses of Parliament, Parliamentary Office of Science and Technology. *POSTNOTE UPDATE NO. 383*. Available at: http://www.parliament.uk/documents/post/postpn_383-carbon-footprint-electricity-generation.pdf. Accessed 05-01-2013.

Rankin, J., 2012. *Energy use in metal production, High Temperature Processing Symposium*, Swinburne University of Technology.

Rankin, J., 2011. *Minerals, Metals and Sustainability: Meeting Future Material Needs*, Collingwood, VIC, Australia: CSIRO Publishing, Australia.

Raugei, M., Frankl, P., 2009. Life cycle impacts and costs of photovoltaic systems: current state of the art and future outlooks. *Energy*, 34, 392–399.

Ray, H. S., Singh, B. P., 2005. Bhattacharjee, S., Misra, V., *Energy in Mineral and Metallurgical Industries*, Allied Publishers PVT LTD, New Delhi, India.

Siddiqi, A., Anadon, L. D., 2011. The water–energy nexus in Middle East and North Africa. *Energy Policy*, 39, 4529-4540.

Van Sark, W. G. J. H. M., Brandsena, G. W., Fleusterb, M., Hekkertc, M. P., 2007. Analysis of the silicon market: will thin films profit? *Energy Policy*, 35, 3121–3125.

US Department Of Energy, 2011. *Critical Materials Strategy*. Washington DC. US.

US Geological Survey. Mineral commodity summaries 2011. Available at http://minerals.usgs.gov/minerals/pubs/mcs/2011/mcs2011.pdf (accessed January 2013).

Wadia, C., Alivisatos, A. P., Kammen, D. M., 2009. Materials availability expands the opportunity for large-scale photovoltaics deployment. *Environmental Science and Technology*, 43, 2072–2077.

Zuser, A. and Rechberger, H., 2011. *Considerations of resource availability in technology development strategies: the case study of photovoltaics, Resources, Conservation and Recycling*, 56, 56–65.

In: Solar Power
Editor: Stephen Bailey
ISBN: 978-1-63321-317-3

Chapter 2

More than Costs: On the Fit between Solar and Renewable Electricity Policy Motivations and Energy System Models

Johan Lilliestam*[*]*, Yann Blumer, Anna Geddes, Mercè Labordena, Leonhard Späth and Oscar van Vliet

Federal Swiss Institute for Technology (ETH)
Institute for Environmental Decision-Making,
Human-Environment Systems, Zürich, Switzerland

Abstract

Solar and other renewable power can play an important role in the future European electricity system, as shown by numerous energy and electricity system model studies. Many of these reports have in common that they focus on cost as the central issue, either by optimizing it or using it as a key variable in simulation. In this paper, we examine the renewable energy policies in Germany, the Netherlands, Spain, Switzerland and the UK in order to identify the policy motivations and constraints for expanding renewables in these countries, in order to find

[*] Corresponding author: Johan Lilliestam, Federal Swiss Institute for Technology (ETH), Institute for Environmental Decision-Making, Human-Environment Systems, Universitätstrasse 22, Room CHN J71, 8092 Zürich, Switzerland. Tel.: +41 44 632 6037; johan.lilliestam@usys.ethz.ch.

out whether this cost-focus reflects actual policy priorities. We find that their renewable energy policies are motivated by climate concerns and international agreements, energy security concerns, as well as job creation, and other macroeconomic or industry policy reasons. They are constrained primarily by short-term costs of renewable electricity support and grid integration challenges. Thus, actual policies are not very concerned with overall system costs and yet this is the issue many researchers focus on in their scenario building. We contend that the political and scientific debates, including scientific system modelling, on deployment of renewable electricity should go beyond mere cost optimisation studies and focus instead on issues that actually drive and constrain policies today.

1. INTRODUCTION

Renewable electricity generation is growing rapidly in Europe, and this growth is expected to continue in the coming decades. The European Commission, for example, expects renewables to grow at least three-fold between now and 2050, even if no renewables policy is enacted ("reference scenario"). With the suitable policies in place – the Commission's "High-RES" scenario – an 800% increase up to 1750 GW of renewables is considered feasible, including 980 GW of wind and 600 GW of solar power (EC 2011b). It appears likely that Europe will end up somewhere between these two extremes.

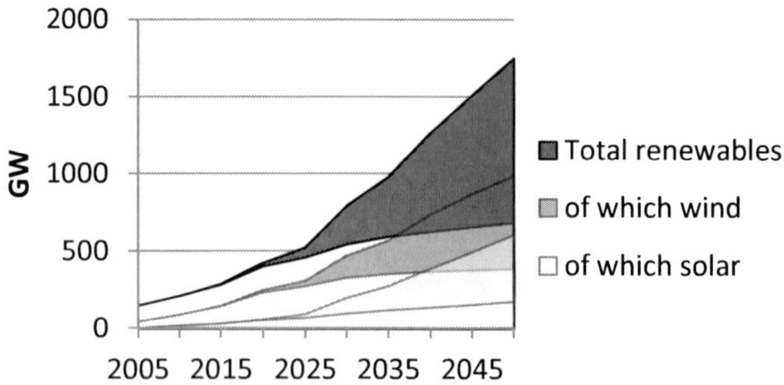

Figure 1. Renewable electricity capacity in the High-RES (upper bound) and Reference (lower bound) scenarios of the European Commission's Energy roadmap 2050. Source: (EC 2011b).

1.1. Modelling the Energy Mix of the Future

The power mix of the future is a topic that has occupied many researchers over recent years, and will continue to do so in the future. Modelling studies by numerous European research and consulting institutions have resulted in thousands of electricity scenarios (Winskel et al. 2014).

An energy model is used either for forward projections or to explore normative scenarios (Pfenninger et al. 2014). A common use of energy models is to assess the effect of (new) goals and constraints placed on the energy system. For example, one can model a new wind electricity policy, both in terms of its effectiveness (e.g. how much wind power can we expect by 2050?) and its efficiency (e.g. how much does that wind power cost?).

Energy models are classified as either a simulation or optimisation type, although the optimisation model is the most common in literature. These models are applied to different scales, from micro-grids of rural communities (e.g. Ashok (2007), Deshmukh and Deshmukh (2008), Nema *et al.* (2009)) to regional electricity systems (e.g. Scholz (2012), Trieb *et al.* (2014)) to global energy system models and integrated assessment models (e.g. Kriegler *et al.* (2014), GEA (2012), Rao *et al.* (2006)).

Two types of optimisation models are distinguished: bottom-up and top-down models. The bottom-up optimisation models exist between engineering and economics. These models are filled with performance and cost data regarding different energy technologies and effectively make these technologies compete given a set of constraints, typically concerning available fossil and renewable resources and (sometimes) emissions. Effectively, they optimise for levelised discounted costs of a particular set of energy services. Well-known examples of bottom-up models are the TIMES/MARKAL and MESSAGE families of energy systems models (IEA-ETSAP 2014, Messner and Strubegger 1995). Bottom-up models do not include macroeconomic effects, though these can be addressed by connecting to other (economic) models (e.g. Messner and Schrattenholzer (2000)).

The top-down optimisation models are closer to econometrics. They use input-output data sets of entire economies, like the Global Trade Analysis Project (GTAP 2014), with modifications based on technology properties, and are exclusively price-driven. Individual energy technologies are described in economic effects, rather than the engineering properties seen in bottom-up models. Top-down models optimise for the largest output of the entire economy (or a similar measure). Well-known examples are the GCAM and GEM-E3 models (Edmonds et al. 2014, EURELECTRIC 2010, JRC 2014).

In principle, these models do the same thing: they generate a normative but possible electricity and/or energy system, optimised for an economic criterion, generally total system cost, while simultaneously fulfilling some specific boundary criteria, typically a long-term climate change mitigation target and/or some system stability rules.

It is possible to add other criteria than cost and technology performance into the optimisation as static input data. For example, monetised consumer preferences and market penetration rates are added into the optimisation in the CIMS model (Jaccard et al. 2004, Mau et al. 2008). Another option to expand the model scope beyond costs is to run the model many times with a large set of permutations of the valuations of non-cost factors and pick a model run that matches the values of the observer. Such an approach, although rare, does allow for a limited multi-criteria analysis (similar to McCollum et al. (2011)).

Simulation models show much more heterogeneity. Whereas optimisation models are naturally bound to a single method, simulation models combine submodules that can be based on different methods like engineering-economic, macroeconomic and agent-based modelling[1]. This allows simulation models to combine the advantages of both top-down and bottom-up optimisation models with insights from other disciplines. However, this flexibility comes at a cost of increased complexity.

In practice, many simulation models favour particular methods, depending on which questions they seek to answer, so that the methodological landscape is rather diverse (see Capros et al. (2014a, 2014b)). A prominent simulation model is PRIMES, which is the primary model used by the European Commission for the construction of its energy scenarios, including the Roadmap 2050 scenarios (EC 2011a). PRIMES is based on micro-economic sub-modules that calculate energy trade, supply and demand by sector (Capros et al. 2014a). Its submodules include engineering data, but less detail than many bottom-up models. Other well-known examples of simulation models are the LEAP and IMAGE-TIMER models (Heaps 2012, Hoogwijk et al. 2007, van Vuuren et al. 2010).

Although the simulation models are more diverse than pure optimisation models, in their core they provide a forecast of how a system will develop, given a set of boundary and policy conditions. These conditions may differ,

[1] All energy models optimise, the difference is that those described as optimisation models optimise as much of the system together as possible, whereas simulation models are deliberately split into submodules that optimise within their own subsystem, and not necessarily simply on price.

but cost is generally a key parameter, and many simulation models have cost-optimising modules (see Pfenninger *et al.* (2014)).

1.2. Model Scope vs. Policy Drivers

Models are simplified representations of reality, given limitations in computation time, data availability, etc. Such simplifications can be done by reducing the resolution in space (e.g. to regional- or global-level descriptions) and/or in time (e.g. to yearly or seasonal averages). Another way to simplify is to reduce the scope, for example by looking only at the cost of a climate-constraint scenario while omitting all other possible energy drivers and constraints. Lack of transparency is a major shortcoming in the development of these modelled scenarios, as energy system models of any kind beyond the trivial contain large amounts of data that are rarely fully published or transparently reviewed (Pfenninger *et al.* (2014); for rare examples of transparency, see OSeMOSYS (2014), van der Sluijs *et al.* (2002)). Yet, what has been modelled and what has not, and hence what the results mean and what they do not mean, are highly important issues for a vastly politicised topic like energy.

Given this opacity and the necessary simplifications used in energy system models, we cannot be certain that they model a) things in the right way and b) the most important factors that actually drive or constrain particular developments in the energy system. Here, we will not investigate how the models arrive at the results, but rather whether their scope covers the main drivers of and constraints to renewable energy policy.

Energy models are not value-neutral: in society, as in policy and science, there are multiple narratives about the suitability of renewables and other energy technologies. This may result in open-ended, inconclusive debates concerning the costs of technologies, with unresolvable disagreement about models' cost assumptions (e.g. Blitz (2010), Kemfert (2013)), or about what should be considered to be "expensive" (Ellenbeck 2013, Munoz *et al.* 2014, Thompson 1984, Zheng and Kammen 2014). Some even reject cost as a parameter of interest for the future energy system, asserting that a switch to renewables is a matter of ethics and a necessity for humankind (e.g. Scheer (2010)). There is a myriad of issues behind the energy debate, such as local environment vs. global climate; justice and equity vs. macro-economic cost benefits; democratisation of energy vs. system cost; efficiency measures vs.

supply-side policy; integration vs. autarky; solar power vs. wind power vs. fossil power, and so on.

These subjective issues are often not immediately visible in the debate. Instead, policy-makers, industry and civil society (especially environmental NGOs) try to objectivise the debate, by relying on science and its models. The cost of a scenario is an intrinsically political question, as is the only seemingly objective question of what is technically feasible. As a consequence, two sides of a debate may both use "scientific evidence" to support their view although this evidence is at conflict: for example, one study may tell us that PV is cheap and nuclear expensive (e.g. Lovins *et al.* (2008)) whereas another tells us the opposite (e.g. Schlesinger *et al.* (2010)). It thus seems that the results of many models, as policy itself, are determined by choices of ideological and political standpoints of how one envisions the future energy system, perhaps more so than by "objective" costs and technical benefits of a particular pathway (Ellenbeck *et al.* 2013, Wynne 1984).

1.3. Identifying Solar Electricity Policy Motivations and Constraints

In this paper, we identify what policy wants to achieve – the motivation – when deciding and implementing national solar electricity policies, and which issues may constrain policy. We examine the motivations and constraints of these policies in 5 European countries. In most countries, solar electricity policy is included in the general renewable electricity policy framework, so that we will look primarily at the justifications of renewable energy policy to identify the motivations and constraints of solar power policy. In some cases, renewables policy is not separate but included in the general electricity or energy policy framework; when necessary, therefore, we will deduce motivations and constraints from general energy policy.

For this work, we seek to answer two questions. First, we want to see what role costs play as a driver for renewable electricity policy in Europe, and how this relates to other motivations. Second, we look at which factors constrain policy, effectively putting boundaries on which renewable electricity futures are feasible from a policy perspective. The resulting array of policy motivations and constraints define the set of optimisation variables and boundary conditions for a holistic, policy-relevant energy model.

We use a simple approach of looking at the policy justifications, as expressed in government statements, supporting material for law propositions, parliamentary debates, etc. In this, we take the government's argumentation – regardless of which parties are in government at a given time – as the official reason for the national policy. We do not take a political scientific view on the policy process (e.g. how policies come to be, whether they are implemented in the intended way, etc.), but rather look only at the policy output and the official, published arguments for this.

We chose the five country cases so as to examine policies that have triggered large-scale expansion of renewable electricity as well as policies that have not, including both mature and still developing public and political renewable energy debates. Therefore, we examine the renewables policy of Germany, as the most successful renewables policy in Europe, along with its highly developed and dynamic public and policy debate. We look at Spain, which was as successful as Germany, before it suddenly suspended its renewables support. We also investigate the policy of the UK, which has applied numerous renewable policy reforms, although none of them has been truly successful, as is the case with the Netherlands. We also study Swiss renewable policy, as a country that has in principle decided to strongly reform its electricity supply structure, but has not yet fully decided how. In this, we look not only at the current (2014) policy, but also how and why the currently existing policy frameworks grew over time, in order to identify the dynamics of renewables policy.

In the following section we describe the solar and renewable electricity policies of the 5 country cases, what these policies were to achieve and the issues that limit solar electricity policy. In section 3, we synthesise our findings from the country cases and discuss the implications for modelling, before we in the last section present our conclusion.

2. Government Motivations and Constraints for Solar and Renewable Energy Policy

2.1. Germany

2.1.1. German Solar and Renewable Electricity Policy

The central element of German solar energy policy is the Renewable Energies Act (Erneuerbare-Energien-Gesetz, EEG). The EEG includes all renewable electricity, not only solar, but there are policy amendments that are solar-specific (e.g. EEG (2012)). This law regulates the priority feed-in, grid connection and technology-specific financial remuneration of renewable electricity generators in the form of a feed-in tariff. The feed-in tariff is reduced by a pre-defined rate at regular intervals (*degression*) to account for learning effects and decreasing costs of renewables. The *difference cost* between the exchange electricity price and the feed-in tariffs paid are split among all eligible electricity consumers, who pay a surplus fee – the *EEG-Umlage* – on their electricity bill. The basis of eligible EEG payers has been substantially diminished over the years in order to protect the domestic industry from excessive electricity costs (EEG 2004, 2008, 2011).

2.1.2. Stated Motivations

The overarching purpose of the EEG is to enable sustainable development in the energy sector (EEG 2000) and to lead Germany into a "renewable era" (EEG 2011). More specifically, German renewable energy policy strives to achieve six specific aims.

The expansion of renewables is primarily justified by national, European and international renewable energy and climate policy objectives (EEG 2000, 2003, 2004, 2008, 2011). Hence, most EEG reforms seek to maintain or increase the speed of renewable electricity expansion, sometimes by increasing the support, in order to achieve these targets. The original objective of the EEG was to at least double the share of renewable energy consumption in Germany by 2010, compared to 2000, by increasing the share of renewable electricity (EEG 2000). This aim was reached in 2006 (BMU 2013). Today, the German objective is to reduce greenhouse gas emissions by 80% by 2050 and to achieve 35% renewable electricity in 2020 and 80% in 2050 (EEG 2011). Currently, renewables satisfy 23.5% of the German electricity demand, up from 6.5% in 2000 (BMU 2013, Eurostat 2014).

A second aim concerns energy security and addresses the long-term availability of energy. German renewables policy refers to renewables as the only way secure long-term energy supply and maintain its political independence, as all other energy forms are exhaustible and finite. Not reducing the dependency on fossil and nuclear energies will lead to an unstable energy supply in the future, at ever increasing costs, and will lead to increased conflicts, or even war, over the remaining fossil and nuclear resources, which are concentrated in only a few, already conflict-laden, world regions (EEG 2000, 2004, 2008, 2011). In 2012, renewable electricity generation displaced 270 TWh of primary coal and 53 TWh primary gas, most of which would have been imported; the total import cost savings were almost 4 billion € (BMU 2013).

The third motivation is related to green growth, and seeks to secure jobs and build up new, sustainable industries. The domestic expansion of renewable energies, it is argued, leads to the creation of a renewable energy industry and new jobs, in particular in sectors in which Germany is traditionally strong, especially mechanical and electrical engineering. Some 380,000 people were employed in the German renewable energy sector in 2012, up from 160,000 in 2004, despite the crash of the domestic PV manufacturing industry (BMU 2013, EEG 2000, 2004, 2008, 2011).

Fourth, the German renewable policy aims to bring down the costs of renewable electricity to the point where it is competitive with conventional electricity in the market (EEG 2000, 2004). Expanding renewables leads to technological learning and economies of scale, which brings costs down – eventually to the point where renewables can stand alone in the market, without a dedicated policy to support them (EEG 2000, 2004, 2008). The full scope of these learning effects can be seen by the massive lowering of feed-in tariffs: the PV tariff in 2000 was at least 46 €c/kWh whereas the tariff in March 2014 was at least 9.2 €c/kWh (Bundesnetzagentur 2014, EEG 2000, 2012). The expansion of renewables is seen as a process that will slightly increase the short-term costs while greatly reducing them in the long term; only in the 2012 solar reform are the short-term costs seen as too high.

Fifth, the environmental – non-climate – effects of renewables are a minor focus in all EEG reforms. An additional aim of expanding renewables is to lower emissions of various pollutants, including SO_2 and NO_x, which cause acid rain, CFCs, and to lower the water consumption of the electricity sector (EEG 2000, 2004, 2008, 2011).

Finally, in the reforms up to 2008, the years in which the Social democrats and/or the Green party were in government, the EEG justifications refer to the risks of nuclear accidents and waste. In the reforms of 2011 and later, when none of these parties were in government, the motivation has changed: renewables are to cover a large share of the capacity lost during the nuclear phase-out, but the reference to nuclear risks is not included (EEG 2000, 2004, 2008, 2011).

2.1.3. Stated Constraints

We found three main issues that constrain the German renewable electricity policy.

The first constraint refers to grid stability and the integration of volatile renewables into the electricity system. This is a focus in the 2011/2012 reforms, but not in the earlier EEG versions. Similarly, the second nuclear phase-out decision in 2011 increased fears of a capacity gap in the future.

The continued renewable electricity expansion is therefore contingent on suitable technical and managerial system integration measures being effective. The continued expansion is also seen as dependent on a sufficient pace in grid expansion in different places in the system, in order to alleviate regional bottlenecks and allow wind power from the coasts in the north to reach the demand centres in the south, and to allow PV electricity fed in at low voltage levels to reach the higher voltage parts of the grid (EEG 2011).

Second, the impacts on local nature and environment are discussed as an important potential constraint, but these are largely seen as easily manageable by complying with the siting rules in the EEG for avoiding local damages. The EEG instead emphasises that renewables are, in sum, much more environmentally friendly than conventional energies (EEG 2000, 2004, 2011).

Third, the costs of supporting renewables are seen as a constraint in 2012 and, partially, in 2011, but not in the earlier EEG versions. In the early versions, renewables are viewed as a means to avoid increasing electricity costs caused by rising fossil fuel prices, and to stabilise electricity prices at a low level in the long run.

In 2004, the reform justification emphasises that renewables are even the cheapest available option for the future energy supply and the cheapest one already today if external costs were included (EEG 2004); the 2000 version states that renewables need support only because the electricity market is distorted, conventional electricity is subsidised and because fossil fuels do not carry their own costs (EEG 2000). In these reforms, the cost of renewable energy support is explicitly discussed as affordable: for example, in 2008, the average household paid 1.10 €/month for the EEG difference cost, whereas it paid 6.50 €/month in 2011 (EEG 2008, 2011).

In 2012, however, the costs of supporting renewables, especially solar PV, are seen as a significant problem. The massive and rapid increase of PV in 2010-2012, following the price decrease in the global market, were seen as the main cause for the rapid EEG cost increase; the *EEG-Umlage* increased to 5.3 €c/kWh in 2013 (BMU 2013).

The principal aim of the 2012 reform was thus to slow the PV expansion. For this, the feed-in tariff was cut and the degression rate increased (EEG 2012). The aim of 80% renewable electricity by 2050 remained in place.

2.2. The Netherlands

2.2.1. Dutch Solar and Renewables Energy Policy

Dutch renewable energy policy, which includes one specific regulation for solar energy, has two pillars. The first pillar consists of three separate subsidy schemes: direct subsidies for small-scale solar energy, tax deduction for investment in renewables, and a feed-in premium that guarantees a minimum revenue for producers (Tweede Kamer 2001, 2007, 2009). The total support expenditures are limited and the subsidies are allocated to cheaper technologies first, and then on a first-come, first-served basis (MinEZ 2012a).

The second pillar consists of *Green Deals* – public-private partnerships, where the Dutch government commits to easing regulatory burdens and other barriers to renewable energy, and a consortium of private partners that commits to building a number of units (MinEZ 2011). A total of 8 Green Deals in solar energy were signed up to March 2014, ranging from applied research to demonstration projects (Ondernemend Groen 2014).

2.2.2. Stated Motivation

The long-term Dutch renewable energy policy aim is to achieve an entirely sustainable energy supply by 2050. The official objective is to attain a 14% share of renewable energy by 2020, of which 40 PJ should come from decentralised sustainable energy (a category that includes solar heat and PV, heat pumps and CHP) (Rijksoverheid 2014).

This expansion of renewables is motivated by three main arguments: fulfilling international obligations, avoiding climate change, and boosting the Dutch economy.

Avoiding climate change and fulfilling international obligations are closely linked, as the Netherlands are bound by EU energy and climate policies and pledges. In practical terms, the quantitative objectives for renewable energy are determined by the EU climate and energy policy package (EC 2014).

In addition, Dutch renewable energy policy is strongly framed as industry policy, as reflected by the strong focus on markets, innovation and public-private partnerships. Unlike many countries, renewables are managed not by the Ministry of the Environment, but by the Ministry of Economic Affairs. The green growth paradigm is particularly visible in the *Energieakkoord* (Energy Pact), an agreement between several dozen companies, NGOs and government departments and agencies at various levels for a transition to a more efficient, clean and renewable energy system. The aim of the *Energieakkoord* is to enable "a cost-effective expansion that offers security for investors, creates additional employment, triggers innovation that reduce costs, and contributes to a strengthened competitive position of Dutch companies in this industry" (SER 2013).

The ultimate goal is that sustainable energy becomes competitive, in the sense that the cost of implementation remains manageable and that a Dutch renewable energy industry emerges. There is a formal ambition that the Dutch CleanTech industry will be in the global top 10 by 2030 (SER 2013). The main focus of this push is the wind power industry, in which the Netherlands has historically been very active. In addition, the *Energieakkoord* also explicitly lists the expected creation of 15,000 new full time jobs that will result from its implementation, in addition to the projected energetic benefits.

2.2.3. Stated Constraints

We found two major policy constraints to the expansion of solar and renewable electricity in the Netherlands. The first is cost of the renewable electricity support, which is demonstrated by the restricted budget of the support schemes, as well as the focus on private investment in the public-private partnerships. The budget for the investment support for home PV systems by private persons is regularly exhausted before the end of the year. Its budget was limited to 21.6 million € in 2012, 51 million € in 2013 and nothing for 2014 (Tweede Kamer 2012a, b, 2013).

Similarly, producers can apply for one of six rounds with progressively higher feed-in premiums of approximately 7, 8, 9, 11, 13 and 15 €c/kWh, but the total budget is also limited (MinEZ 2012a). In the long run, the government expects costs of renewable energy technologies to come down, and prices of fossil energy to go up, making renewables competitive in the long term, thereby eliminating the need for subsidies (Rijksoverheid 2014).

The feed-in scheme allowed for 3 billion € of support commitments in 2013, a budget which was raised to 3.5 billion in 2014 (MinEZ 2012b, 2013). This makes it "the most important instrument for financing [...] the expansion of renewable energy production" (SER 2013).

A second issue is that policy considers the Dutch energy system to be unprepared for a large increase in renewable electricity. The government sees this largely as a regulatory issue, and an impending overhaul of the electricity law will impel relevant business, such as transmission and distribution grid companies, to adequately adapt their infrastructure (MinEZ 2014).

2.3. Spain

2.3.1. Spanish Solar and Renewable Electricity Policy

The foundation for solar and renewable electricity deployment in Spain is the General Electricity Law 54/1997 (Ley 54/1997 1997). The law, which has been modified repeatedly, created a *special regime* for electricity from renewable sources and for cogeneration, under which the eligible generators receive unique tariffs and guaranteed grid access for qualified electricity production. The feed-in tariff and market premium incentives established in the mid-2000s to promote solar PV were some of the most generous incentives at that time. As a reaction to fiscal and economic difficulties faced during the financial crisis, however, the government decided to greatly reduce the renewable electricity support in 2010 (RDL 2010a, b). In January 2012, the government suspended the feed-in tariffs and market premiums incentives to all new renewable electricity installations (RDL 2012). In July 2013, the incentive scheme for renewables was cancelled and incentives were replaced by subsidies based on the rate of return on Spanish treasury bonds (RDL 2013).

2.3.2. Stated Motivations

The first motivation for the Spanish renewable energy policy is its international and European climate and renewable energy commitments. The Spanish renewable electricity targets are transposed from the European Renewables Directive: 20.8% of renewable energy and 38.1% renewable electricity by 2020 (REP 2011). The high financial incentives for renewable electricity resulted in a sharp increase in investment, so that most technologies exceeded the government targets: for example, PV capacity exceeded the 2010 400 MW target more than tenfold and solar thermal power installations were three times above the 500 MW target (NREAP 2010, REP 2005). Despite the rapid deployment of renewables, the Spanish greenhouse gas emissions increased during the first Kyoto period, and it is not on track for meeting its 2020 target of -10% emissions compared to 2005 (EEA 2013).

The second aim of the Spanish renewable energy policy is to decrease the dependence on energy imports, which is presently around 80%. This high import dependence is costly for the Spanish economy and exposes Spain to the price volatility of international energy markets. In particular, energy imports are a significant contributor to the negative Spanish balance of trade. Hence, to reduce the energy import dependency and improve the balance of trade, the REP aims to increase the share of domestic renewable energies (REP 2011).

The third motivation is to enhance the competitiveness of the Spanish national economy. The expansion of renewables, and especially solar power, will contribute to cost reductions, eventually achieving grid parity, ultimately making it a long-term low-cost option for consumers (REP 2011). The cost-reducing effect of the solar energy expansion in Spain and globally can be seen in the continuous lowering of the Spanish feed-in tariffs: the PV tariffs in 2004 were up to 40 €c/kWh, or 575% higher than the average electricity price (RD 2004). Six years later, the tariffs were 13-28 €c/kWh, compared with the 16 €c/kWh average consumer price (RDL 2010a). The reforms before 2010 viewed the expansion of solar power and renewables in general as a process that marginally increases the short-term costs while decreasing the medium- to long-term costs. Yet, in the reform of 2012 (RDL 1/2012), short-term costs of supporting renewables were seen as too high and financial incentives were suppressed (see 2.3.3).

The fourth aim is to secure a sustainable economic, social and environmental national development. This green growth objective especially translates into triggering investments and creating new, highly qualified and sustainable jobs. In 2009, solar photovoltaic was the renewable technology that directly contributed the most to GDP (2,789 million €), followed by wind (1,993 million €) (REP 2011). In 2010, the production, installation and operation of renewable energy facilities in Spain had created 70,000 direct jobs, of which 28,000 were in the solar PV business, and another 46,000 indirect jobs (REP 2011). Although the official plans were to use mainly Spanish-produced components, the rapid PV expansion could not be satisfied by locally manufactured modules. Hence, in 2008 – the year with the highest number of solar PV installations – Spain imported 2.8 billion € worth of solar components. Following the suspension and later cessation of the financial support for renewables (RDL 2012, 2013), about 90% of the PV industry went out of business; between 2010 and 2013, around 24,000 of the 28,000 PV-related jobs were lost (UNEF 2013).

2.3.3. Stated Constraints

We found three main constraints to solar and renewable electricity policy in Spain. First, the support costs of renewable energy are the most significant constraint to Spanish renewable electricity policy. Up to 2010, the future benefits of renewables were seen as outweighing the short-term costs, thus justifying the economic support scheme; in 2010, this changed and the support for renewable electricity was openly stated to be too expensive (RDL 2010a, b, REP 2011). The generous support schemes triggered a rapid renewables expansion, and as a consequence, the costs of renewable energy support soared, contributing to a strongly increasing *tariff deficit*. The tariff deficit describes the gap between revenues collected from consumers – the price-change rate for which is capped – and the system costs, including support for renewables but also, for example, non-renewable generation for islands. The tariff deficit is paid with public funds, but following the financial crisis, the government had neither the cash flow to cover this cost nor the flexibility to make fast pricing adjustments. Hence, the Spanish renewable electricity support system was greatly reduced in 2010 in order to limit the deployment of especially PV (RDL 2010a, b), and was cancelled in 2013 (RDL 2013). Today, although renewables support has been cancelled, the tariff deficit has continued to increase to more than 26 billion € (RDL 2013).

Second, the limited transmission interconnection capacity with neighbouring countries is seen as an important constraint to the expansion of renewables. This limited capacity restricts energy exchanges with neighbouring countries to only 3% of internal consumption – the lowest interconnection level in Europe. This limits the expansion of volatile renewables, as electricity cannot be exported in significant amounts in times of surplus or imported in times of shortage. This is particularly the case in the wake of the economic recession, which reduced the electricity demand: consequentially the power system may be oversupplied in the near future, unless the interconnections are expanded (REP 2011).

Third, grid stability and integration of volatile energy into the electricity grid are seen as a major constraint for the expansion of renewables. This technical stability of the electricity system first appears as a serious concern in 2010 (RDL 2010a, b), after the share of wind and solar PV capacity amounted to over 25% of all installed capacity (Eurostat 2013, REE 2013). Further, the rapid growth of renewable electricity was not matched with a parallel increase in transmission and distribution grid capacity, leading to increasing strain on the grids (see above). As a result, the government settled a toll of 0.05 €c/kWh to all electricity generators for electricity fed into the grid and increased the grid integration requirements for PV installations exceeding 2 MW (RDL 2010a, b). Following the recent calls of the European Commission to develop strategies to ‘empower consumers’ and encourage demand-controlled production of renewables, Spain authorized self-consumption of electricity (without any specific premium) to small business and the domestic sector for systems up to 100 kW (Ley 24/2013 2013, RD 2011). Self-consumption aims to ensure the deployment of PV after it has reached grid parity by cutting all financial supporting costs and facilitate the transition towards a post feed-in tariff electricity future. Discussions are still ongoing about the introduction of a net-metering scheme, in which the compensation of electricity generation would be calculated on a yearly basis while some grid costs may require to be paid by the photovoltaic system owner.

2.4. Switzerland

2.4.1. Swiss Solar and Renewable Electricity Policy

In the absence of a dedicated renewable electricity policy, the deployment of renewables in Switzerland is regulated under the general national energy policy. The Energy Act (Loi sur l'énergie, LEne), which was released in 1998, adapted in 2007, and is currently (March 2014) being reworked, is the legal basis for the solar and renewable electricity policy in Switzerland (Swiss Federal Parliament 2014). Since 2009, the main instrument to develop renewable electricity is a feed-in tariff, which is financed through a charge added to the electricity grid fee for small and medium-size consumers.

2.4.2. Stated Motivations

The LEne aims to ensure that the energy supply of Switzerland is cost-competitive and environmentally sustainable, is geared towards minimal consumption, that prices reflect the true cost of energy provision and that the share of domestic and renewable energy resources is expanded. The current revision of the LEne seeks to pave the way for the Strategy 2050 of the Federal Council while adhering to these principles (Swiss Federal Council 2013a). Some of the motivations for promoting renewables are neither explicitly nor consistently defined in the numerous governmental documents that compose Switzerland's energy policy. Yet, a number of motivations can be identified, with varying degrees of detail.

First, the Federal Council aims to reduce CO_2-emissions from around 5 tons/capita per year today to 1-1.5 tons/capita per year by 2050 (Swiss Federal Council 2013a). For the already virtually carbon-neutral electricity sector (BFE 2013d), this target means that renewable electricity should account for any future demand increases and compensate capacities being decommissioned. Climate change mitigation can be considered the most important driver for the introduction of the feed-in tariffs for renewable electricity in 2009.

Second, following the Fukushima nuclear disaster, Switzerland has decided to phase out nuclear power. The ongoing revision of the LEne thus aims to fill the resulting capacity gap as far as possible with renewable electricity, as renewables are the only domestically available energy resource (BFE 2013a, Prognos 2012). The five existing nuclear reactors are allowed to continue operations but will not be replaced on reaching the end of their planned operational life of 50 years. The first reactor will be decommissioned in 2019 and the last in 2034.

Third, the LEne aims to promote domestic energy. While this is not possible with respect to oil and gas, Switzerland is currently able to cover its domestic electricity demand by hydropower and nuclear power. Increasing electricity demand and the nuclear phase-out have triggered fears of an *energy gap*, i.e. a difference between projected consumption and production, which is to be reduced as far as possible by expanding domestic renewables (Swiss Federal Council 2013a).

These aims have led to the quantification of specific targets for the expansion of renewables. By 2030, the annual renewable electricity production (including small-scale hydropower) should increase by 5.4 TWh from the 2000 levels (Swiss Federal Council 2013a). This represents 8% of the electricity generation of 2012, and would compensate less than ¼ of the nuclear power generation (BFE 2013c). Currently, the share of new renewable electricity (including wind, PV and biogas power, but excluding hydro power) is less than 1% (BFE 2013d).

2.4.3. Stated Constraints

The Swiss expansion of renewable electricity is primarily constrained by the short-term costs of renewable electricity support, which are limited both for each year and in the long term. Currently, the short-term costs to consumers of the feed-in tariff is capped at 0.009 CHF/kWh (about 0.7 €c/kWh), up from 0.006 CHF/kWh in 2012 (Swissgrid 2014). However, this scheme triggered more projects than expected and the budget limit has been prematurely reached, causing a blockade of more than 30,000 renewable energy projects (Swiss Federal Council 2013a). In the current version of the LEne, the support costs are allowed to increase to a maximum of 0.014 CHF/kWh, about 1.1 €c/kWh (Swiss Federal Parliament 2014). If renewable electricity expands too much – so that the support costs exceed that limit – additional projects will go without support. Hence, although the limit can be changed in future reforms, "too high" support costs are presently defined by law.

Other issues, like grid stability, are not explicitly expressed in the policy justifications as constraining factors for the expansion of renewable electricity in Switzerland. However, in the newest national strategy on electricity grids (Swiss Federal Council 2013b), the potential challenges of intermittent renewables to the stability of the grid are referred to and several potential solutions are outlined, such as shutting down renewable generation in times of overproduction. In general, however, renewables – as the only domestic energy resource – are rather seen as an energy security solution than as a problem.

Environmental constraints, especially the preservation of local ecosystems threatened by the expansion of hydro power, play an important role in several policy documents, but are not explicitly discussed in the justifications for existing or proposed legislation (BFE 2013b, Swiss Federal Council 2013a).

2.5. The United Kingdom

2.5.1. UK Solar and Renewable Electricity Policy

Since market liberalisation and privatisation in the 1980s and 1990s, the UK has seen a strong increase in energy and climate change policy creation. A diverse collection of parliamentary acts, energy white papers, energy reviews, plans, strategies and roadmaps form the overarching energy policy framework, currently under the remit of the Department of Energy and Climate Change (DECC). Solar and renewable electricity policy is embedded in this framework (CCA 2008, DECC 2009, 2011, 2012, 2013b, c, DTI 2003, 2006, 2007, EA 2013).

With the approval of the Climate Change Act 2008, the UK became the first country to set a legally binding, long-term carbon reduction target into law (CCA 2008). Hence, in addition to implementing the EU Renewables Directive, UK government ministers have a statutory duty to introduce and maintain emissions reduction policies. It should be noted however that missing these targets merely results in a requirement for the government to show parliament how they will subsequently compensate for the excess emissions. More recently the 2011 Energy White Paper and the Energy Act 2013 introduced Electricity Market Reform (EMR) (DECC 2011, EA 2013). The EMR is to provide long-term support for all types of low carbon generation and thereby diversify the energy mix by removing exposure to volatile wholesale electricity prices (DECC 2011, EA 2013). In addition to planning and permitting reforms, three financial support schemes aim to trigger investments in renewables: Contracts for Difference (CfD) scheme, which replaces the current Renewables Obligation system, is to encourage large-scale solar PV, a feed-in tariff system aims to encourage small-scale PV, and the introduction of a carbon price floor (CPF) is to trigger investment in low-carbon technologies in general (DECC 2011).

2.5.2. Stated Motivations

The solar and renewable electricity policy of the UK is driven by a set of three main motivations. First, ensuring reliable future supplies is the key objective of the UK energy policy, including its renewables policy (DECC 2011). As the British domestic fossil fuel resources are being depleted, continuing down a fossil-based energy path would increase the UK's dependency on energy imports, making it vulnerable to political events outside the UK. As renewable energy is available domestically, increasing its share in the electricity supply would relatively reduce the need for energy imports. In addition, both the power plant fleet and the electricity grids are rapidly aging, and large shares of the capacity must be replaced – to a large part by renewables – in the short- to medium term (DECC 2011).

Second, the EU Renewables Directive commits the UK to a 15.4% share of renewables in the final energy consumption in 2020. To achieve this, the UK has defined a target of 30% of electricity demand to be met by electricity generated from renewable energy sources by 2020; in 2011, the renewable electricity share was 8.7%, up from 4.5% in 2006, whereas renewable energy accounted for 4.1% of energy consumption in 2013, up from 1.6% in 2006 (DECC 2013a, DUKES 2013, Eurostat 2013).

Third, UK policy strongly emphasises reducing carbon dioxide emissions as an important aim, and has a legally binding target of reducing greenhouse gas emissions by 80% by 2050 compared to 1990. To ensure that regular progress is made towards this long-term target, the Climate Change Act also established a system of five-yearly carbon budgets. The first four carbon budgets, leading to 2027, have been set in law and require that emissions be reduced by 50% on 1990 levels in 2025 (CCA 2008, CCC 2014). The UK is currently in the second carbon budget period (2013-17) and is on track to outperform the first two carbon budgets; this is however mostly because of the impact of the financial crisis. Meeting future carbon budgets will require reducing emissions by at least 3% a year, and less than 1% was achieved in the first non-recession year, 2011. Hence, UK renewables policy aims to achieve a faster expansion of renewable electricity generation than in the past, in order to reach the future emissions reduction targets (CCC 2014).

2.5.3. Stated Constraints

Current UK renewable energy policy is constrained by two main factors. A key concern is grid stability, given the intermittent nature of some renewables. The UK electricity grid and associated infrastructure is aging and is not considered able to provide a reliable electricity supply as the share of volatile renewables increases. The Energy White Paper 2011 states that "vast" investment is needed to upgrade the grid to allow further expansion of renewables whilst also providing a stable supply of electricity. It also concedes that the electricity market in its current form will not be able to provide this required investment. Given that security of future electricity supply is a key policy objective, these concerns *de facto* act as a significant constraint to expansion of renewable energy technologies (DECC 2011).

Although policymakers have acknowledged that some renewable technologies are currently more expensive than traditional fossil fuel generation, the issue of their cost is not explicitly stated as a constraint to their further expansion. A key objective in the general energy policy, however, is to keep energy bills affordable, without further specifying what is affordable. The current energy policy framework states that costs to consumers must be minimised, and that investments must take place in the most cost-effective way possible (DECC 2011, 2013b, c). Given this general energy policy aim, the cost of renewables – specifically the cost of renewable energy support – should be seen as a potential policy constraint in the UK.

3. Discussion

The five European policy cases we assessed are similar in their motivations and perceived constraints for solar and renewable energy policy, but differ greatly both in breadth and depth concerning their argumentation. In the next two sections, we summarise the motivations (section 3.1) and constraints (section 3.2), before we discuss what this means for modelling in section 3.3.

3.1. Motivations

All countries find the reason for expanding renewables, including solar power, in at least two of six categories of arguments, see Table 1. The renewable energy policies in all investigated countries are motivated by climate change mitigation. In addition to a national strive for climate protection, the pace and level of the expansion of renewables is justified by reference to international commitments. Further, the EU member states all refer to the binding European renewable energy targets, which in turn are motivated by a set of issues, primarily climate protection. Four countries are motivated by long-term energy security concerns by securing access to domestic energy and reducing import dependency. Switzerland and Germany view renewables as a domestic way to replace nuclear power. Green growth, prominently the creation of new jobs and a renewable energy industry, is a focus of German, Dutch and Spanish renewable energy policy; Germany and Spain also justify their renewable energy policies by lower electricity prices in the long run.

Table 1. Summary of stated renewable and/or solar energy policy aims in the five country cases

	Climate change	European renewable energy targets	Nuclear phase-out	Long term energy security (import dependency, depletion)	Green growth/ compete-tiveness (jobs, growth...)	Long term, low cost electricity
Germany	x	x	x	x	x	x
Spain	x	x		x	x	x
Netherlands	x	x			x	
Switzerland	x		x	x		
UK	x	x		x		

3.2. Constraints

We could identify four main issues that effectively constrain renewable electricity policy. Each country policy mentions at least one of these constraints. The costs of supporting still uncompetitive technologies are seen as a primary constraint for renewables in all countries (see below for further discussion). Maintaining the stability of the grid and the integration of volatile generation forms a key constraint in all cases: renewables, yes, but only as much as the system can take. Partially related to grid stability, Spain views a potential oversupply of electricity as a constraint. Germany and Switzerland mention local environment impacts as potential constraints to renewable electricity, see Table 2.

Table 2. Summary of stated renewable and/or solar energy policy constraints in the five country cases

	Short-term energy security (grid stability)	Costs of renewable electricity support scheme	Oversupply of electricity	Local environmental impacts
Germany	x	x		x
Spain	x	x	x	
Netherlands	x	x		
Switzerland	x	x		x
UK	x	x		

The short-term costs of renewables – especially the cost of the renewable energy support – are seen as a constraint in all cases. All five policies explicitly or implicitly (e.g. by giving specific cost examples) state that they are primarily worried about consumer costs – the added "renewables fee" on the electricity bill, which pays for the national support scheme – being too high. In most cases, the support cost constraint is discussed in vague terms, typically by referring to electricity having to remain "affordable", without introducing a hard cost limit or operational definition of when costs are "too high". In Switzerland, however, this point is defined by law to be 0.014 CHF/kWh (1.1 €c/kWh). The UK policy specifies that it wants to minimise consumer costs, but without specifying more precisely what that means; the Netherlands also do not define a "too high" level, but the overall support budget is capped.

The German and Spanish cases stand out, as the short-term cost constraint was binding in these two countries: both countries have acted to reduce the speed and costs of the solar power expansion, which was perceived as too expensive. None of the other countries have done this.

In both the German and Spanish cases, the support cost was repeatedly mentioned by policy as a potential problem in all policy reforms, but it was explicitly dismissed as not being an actual problem before 2010 (Spain) and 2011 (Germany).

The Spanish policy was stopped when the support reached 2.4€c/kWh[2], whereas the Germans experienced 3.5 €c/kWh when PV support was cut in 2011 and the ongoing 2013/14 reform was triggered by a support level of 6.2 €c/kWh in 2014 (BMU 2013). Thus, the level at which support is deemed too expensive varies strongly (from 1.1 €c/kWh in Switzerland to 3.5 or 6.2 €c/kWh in Germany), indicating that the perception of the support cost may be more important than its actual level; this level can also be case- or country-specific.

Additionally, the Spanish and German policy domains show an interesting tension between the aim of low long-term costs of renewables (through learning effects for renewables and price-increases for conventional energy) and the constraint of high short-term costs. This further highlights the subjective and argumentative nature of the term: the term "cost" can be used to describe very different things, depending on what one wants to achieve.

Similar to the support cost constraint, none of the countries defines a level of sufficient grid stability. It appears to be a more serious issue in Germany and Spain, which experienced rapid PV and wind expansions, than in the other countries, which experienced much slower expansions (or none at all). This appears logical: grid integration is more of a problem in Germany, which has 32% intermittent generation capacity, than in the UK, which has 8%. In Switzerland, which has less than 1% PV and wind power capacity, grid integration is mentioned but is far from a focal point of renewable energy policy (BFE 2013d, Eurostat 2013). In no country has the renewable electricity expansion been stopped or deliberately slowed down because of grid integration problems; whether and at what level of intermittent generation this constraint may become binding remains to be seen.

3.3. Implications for Modelling

For the modelling of energy scenarios, these findings have a number of implications.

First, optimising cost of a carbon-constrained energy system is consistent with a narrow subset of prevailing policy aims and constraints, as climate

[2] This number was calculated based on the total support (6876 million €) divided by the total Spanish electricity consumption (291.8 TWh) in 2011. Hence, this number is not entirely comparable to the German number, as the EEG support is not carried by all consumers. The same calculation for Germany (2011) gives 2011 costs of 2.2 €c/kWh, compared to the EEG-Umlage of 3.5 €c/kWh in that year (BMU 2013, CNE 2012, Eurostat 2013).

protection is a primary driver and costs are a barrier in all cases. Nevertheless, one needs to be precise with what "costs" mean: cost of what, for whom, in what perspective?

In the investigated policies, short-term costs of renewables support are an important policy constraint, and one that has been binding in at least two cases. All energy system models could, in principle, model this type of cost and even optimise the system for this parameter, but it is rarely done. The total discounted system cost, which is optimised in bottom-up models, is only an indirect constraint to the policies we investigated. The fit in scope between energy system optimisation models and observed policy motivations is thus limited. Yet, even if the short-term costs were optimised in a model, such a modelling exercise would be relevant but still incomplete, as a number of important drivers and constraints are not explicitly assessed.

Job-creation is a possible factor that can be included in modelling (e.g. ASTRA, see Jochem et al. (2008)), but is not used as an optimisation variable in most bottom-up models. Top-down optimisation models capture the macroeconomic costs of renewables, but they do not explicitly include the macroeconomic benefits – like new jobs – in the optimisation. Yet job-creation and renewable energy policy as a general industry policy measure are key drivers of renewables policy in many countries. Explicit figures on net (or at least gross) job-creation in scenarios are thus highly relevant information for policy and are key outputs from an energy system model. Conversely, a model optimising a carbon-constrained energy system for job-creation while informing about the costs of this would be highly interesting for science and policy alike.

Second, long-term energy security – secure access to energy, especially domestic energy, in the long run – is a main policy motivation for renewables. Depletion risks are generally included in large-scale optimisation models in the form of global constraints of finite fuel availability, but assume that remaining resources are distributed according to economic, not political, rules. If this were true, rich European countries would not need to worry about energy access. However, being shut out or cut off from global energy sources is precisely what they fear (see also Lilliestam (2013)).

Hence, although depletion is included in most models, the most important geopolitical energy security aspect – import dependency and the political risks that come with it – is not included and it is very rarely, if ever, optimised. Occasionally, import dependency (of region X on region Y) is presented as an output of a model (e.g. GEA (2012)), but because of the low geographical resolution, such results often do not give the required information. The dependence of, say, Spain on Algerian gas would be a result of interest, whereas the European dependence on Middle Eastern and North African gas is less important: if problems were to occur, they are most likely to happen between countries, but not between largely arbitrary bundles of geographically proximate states.

Third, short-term energy security – grid integration of volatile renewable electricity generation – is a very important constraint to renewable electricity policy. It is included in some simulation models and bottom-up models looking at national or lower levels, but only very coarsely in regional and global bottom-up models and not at all (at least not explicitly) in top-down models. In many cases, the formalisation is too coarse – either in terms of sectoral scope, or temporal or spatial resolution – to give insights into the stability of an actual electricity system. Yet, system stability appears to be a hard constraint for the expansion of renewables, and thus among the most important issues to assess. If volatile renewables cause blackouts in the future, policy will have a very hard time justifying their continued expansion, especially if they still need economic support.

Fourth, some highly relevant issues, like landscape impacts and acceptance of nuclear power (as reflected in a phase-out decision) are local or national. Optimisation models, in contrast, optimise globally (in a system sense, not necessarily in terms of geography), for all countries in their scope, which means that such local concerns cannot be adequately included. For example, an optimisation model result showing an expansion of nuclear power in Europe may not be very relevant for Switzerland and Germany, which have decided that they do not want nuclear power, despite any possible cost advantages. France and Finland, on the other hand, seem determined to pursue nuclear power almost regardless of the cost. This highlights that cost is not always a primary determinant for policy choices, which instead are based on other factors and may differ strongly even between neighbouring countries.

Conclusion

Our findings indicate that it is both meaningful and interesting to optimise or simulate the levelised discounted system costs of a carbon-constrained energy future, but it has relatively little to do with what policy aims to achieve. In particular, more policy-relevant modelling exercises should focus on optimising or simulating factors other than system cost. Our analysis of five European countries has shown that system stability and grid integration as well as the short-term costs of renewable energy support are the most important constraints, whereas climate protection, long-term energy access, reduced import dependency and job-creation are key motivations. At least these 6 elements thus need to be elements of a holistic, policy-relevant energy system model. We are not aware of any existing model explicitly including all these factors. Hence, to improve their relevance outside the scientific realm, future energy models and the assessments derived from them should be constructed to explicitly include, and possibly use as optimisation variables, the motivations and constraints of actual energy policies.

REFERENCES

Ashok, S. (2007). "Optimised model for community-based hybrid energy system." *Renewable Energy 32*: 1155-1164.

BFE (2013a). Energieperspektiven 2050 - Zusammenfassung. Ittigen, Swiss Federal Office of Energy (BFE).

BFE (2013b). Perspektiven für die Grosswasserkraft in der Schweiz. Ittigen, Swiss Federal Office of Energy (BFE).

BFE (2013c). Schweizerische Elektrizitätsstatistik 2012. Ittigen, Swiss Federal Office of Energy (BFE).

BFE (2013d). Schweizerische Gesamtenergiestatistik 2012. Bern, Swiss Federal Office of Energy (BFE).

Blitz, M. (2010). "Die Menschen wollen keine Atomkraft." *Schrägstrich*: 4-7.

BMU (2013). Erneuerbare Energien in Zahlen. Stand Juni 2013. Berlin, Bundesministerium für Umwelt, Naturschutz und Reaktorsicherheit (BMU).

Bundesnetzagentur (2014). Photovoltaikanlagen: Datenmeldungen sowie EEG-Vergütungssätze. Retrieved 2014.03.03, from http://www.bundesnetzagentur.de/cln_1911/DE/Sachgebiete/ElektrizitaetundGas/Unternehmen_Institutionen/ErneuerbareEnergien/Photovoltaik/DatenMeldgn_EEG-VergSaetze/DatenMeldgn_EEG-VergSaetze_node.html;jsessionid=4F38C19D8AD37970C413EE804E639D01 - doc405794bodyText4.

Capros, P., L. Paroussos, P. Fragkos, S. Tsani, B. Boitier, F. Wagner, S. Busch, G. Resch, M. Blesl and J. Bollen (2014a). "Description of models and scenarios used to assess European decarbonisation pathways." *Energy strategy reviews 2*(3-4): 220-230.

Capros, P., L. Paroussos, P. Fragkos, S. Tsani, B. Boitier, F. Wagner, S. Busch, G. Resch, M. Blesl and J. Bollen (2014b). "European decarbonisation pathways under alternative technological and policy choices: a multi-model analysis." *Energy strategy reviews* 2(3-4): 231-245.

CCA (2008). Climate Change Act 2008 Chapter 27. London, UK Parliament.

CCC (2014). *How the UK is progressing.* Retrieved 6 March, 2014, from http://www.theccc.org.uk/tackling-climate-change/reducing-carbon-emissions/how-the-uk-is-progressing/.

CNE (2012). *Informe sobre el sector energetico Español. Parte I.* Medidas para garantizar la sostenibilidad económico-financiera del sistema eléctrico. Madrid, Comisión nacional de energía (CNE).

DECC (2009). *The UK Renewable Energy Strategy 2009.* London, Department of Energy and Climate Change.

DECC (2011). *Energy White Paper: Planning our electric future - A White Paper for secure, affordable and low-carbon electricity.* London, Department of Energy and Climate Change.

DECC (2012). *UK Renewable Energy Roadmap Update.* London, Department of Energy and Climate Change.

DECC (2013a). *Average Annual Domestic Electricity Bills by Home and Non-Home Supplier* (QEP 2.2.1). Retrieved 17 March, 2014, from https://http://www.gov.uk/government/uploads/system/uploads/attachment_data/file/267329/qep221.xls.

DECC (2013b). *UK Renewable Energy Roadmap Update. London,* Department of Energy and Climate Change.

DECC (2013c). *UK solar PV strategy part 1: Roadmap to a brigher future.* London, Department of Energy and Climate Change.

Deshmukh, M. K. and S. S. Deshmukh (2008). "Modeling of hybrid renewable energy systems." *Renewable and Sustainable Energy Reviews 12*: 235-249.

DTI (2003). *Energy White Paper: Our energy future - creating a low carbon economy.* London, Department of Trade and Industry.

DTI (2006). *The Energy Challenge: Energy Review Report.* London, Department of Trade and Industry.

DTI (2007*). Meeting the Energy Challenge: A White Paper on Energy.* London, Department of Trade and Industry.

DUKES (2013). *Digest of United Kingdom Energy Statistics 2013.* London, Department of Energy and Climate Change.

EA (2013). *Energy Act 2013* Chapter 32. London, UK Parliament.

EC (2011a). *Energy roadmap 2050.* Impact assessment and scenario analysis. Brussels, European Commission.

EC (2011b). *Impact assessment. Accompanying the document Energy roadmap 2050.* SEC(2011)1565. Brussels, European Commisison (EC).

EC (2014). *The 2020 Energy and Climate Package.* Brussels, European Commission (EC).

Edmonds, J. A., J. M. Reilly, S. J. Smith, P. Patel, K. Calvin, M. Wise and J. Eom (2014). *"GCAM wiki."*

EEA (2013). *Trends and projections in Europe 2013.* Copenhagen, European Environment Agency (EEA).

EEG (2000). *Gesetz für den Vorrang Erneuerbarer Energien (Erneuerbare-Energien-Gesetz* -- EEG). Endgültige Fassung. Berlin, Deutscher Bundestag.

EEG (2003). *Zweites Gesetz zur Änderung des Erneuerbare-Energien-Gesetzes vom 22. Dezember 2003.* Bonn, Bundesgesetzblatt I 2003, 68.

EEG (2004). *Konsolidierte Fassung der Begründung für den Vorrang Erneuerbaren Energien* (Erneuerbare-Energien-Gesetz -- EEG) vom 21. Juli 2004. Bonn, Bundesgesetzblatt. 2004 I. S.1918.

EEG (2008). *Begründung zu dem Gesetz für den Vorrang Erneuerbarer Energien* (Erneuerbare-Energien-Gesetz). Konsolidierte Fassung. Berlin, German Federal Ministry for the Environment.

EEG (2011). *Gesetzentwurf eines Gesetzes zur Neuregelung des Rechtsrahmens für die Förderung der Stromerzeugung aus erneuerbaren Energien.* Berlin, German Federal Government.

EEG (2012). *Entwurf eines Gesetzes zur Änderung des Rechtsrahmens für Strom aus solarer Strahlungsenergie und zu weiteren Änderungen im Recht der erneuerbaren Energien.* Berlin, German Parliament.

Ellenbeck, S. (2013). *The social construction of costs in macro-economic studies.* Paris, European Association for Evolutionary Political Economy (EAEPE).

Ellenbeck, S., P. Schmidt, A. Battaglini and J. Lilliestam (2013). "Der Strommarkt als soziale Institution - Einer erweiterte Perspektive auf die deutsche Diskussion um Kapazitätsmechanismen." *Vierteljahreshefte zur Wirtschaftsforschung 03.2013*: 171-182.

EURELECTRIC (2010). *Power choices. Pathways to carbon-neutral electricity in Europe by 2050.* Brussels, EURELECTRIC.

Eurostat (2013). *EU energy in figures.* Luxembourg, Eurostat.

Eurostat (2014). *Greenhouse gas emissions.* Retrieved 2014.03.05, from http://appsso.eurostat.ec.europa.eu/nui/show.do?dataset=env_air_gge&lang=en.

GEA (2012). *Global energy assessment - Toward a sustainable future.* Cambridge, New York, Laxenburg, Cambridge University Press, International Institute for Applied Systems Analysis (IIASA).

GTAP (2014). *Global Trade Analysis Project (GTAP).* Retrieved 2014.04.08, from https://http://www.gtap.agecon.purdue.edu/default.asp.

Heaps, C. G. (2012). *Long-range Energy Alternatives Planning (LEAP) system.* Retrieved 2014.04.03, from http://www.energycommunity.org/ www.energycommunity.org.

Hoogwijk, M., D. P. van Vuuren, B. J. M. de Vries and W. Turkenburg (2007). "Exploring the impact on cost and electricity production of high penetration levels of intermittent electricity in OECD Europe and the USA, results for wind energy." *Energy 32*: 1381-1402.

IEA-ETSAP (2014). "TIMES documentation."

Jaccard, M., R. Murphy and N. Rivers (2004). "Energy–environment policy modeling of endogenous technological change with personal vehicles: combining top-down and bottom-up methods." *Ecological Economics 51*: 31-46.

Jochem, E., C. C. Jaeger, A. Battaglini, W. Eichhammer, F. Idrissova, J. Köhler, D. Köwener, J. Krause, W. Lass, J. Lilliestam, F. Meissner, M. Ragwitz, M. Rauschen, F. Reitze, W. Schade, F. Sensfuß, F. Toro, R. Walz and e. al. (2008). *Investitionen für ein klimafreundliches Deutschland.* Potsdam, Karlsruhe, Potsdam-Institut für Klimafolgen forschung, Fraunhofer-Institut für Innovations- und Systemforschung.

JRC (2014). GEM-E3 model. Retrieved 2014.04.08, from http://ipts.jrc.ec. europa.eu/activities/energy-and-transport/gem-e3/model. cfm.

Kemfert, C. (2013). *Kemfert: EU nutzt überholte Energieszenarien.* Retrieved 2014.04.01, from http://www.dw.de/kemfert-eu-nutzt-%C3%BCberholte-energieszenarien/a-16994313.

Kriegler, E., J. P. Weyant, G. J. Blanford, V. Krey, L. Clarke, J. Edmonds, A. Fawcett, G. Luderer, K. Riahi, R. Richels, S. K. Rose, M. Tavoni and D. P. Vuuren (2014). "The role of technology for achieving climate policy objectives: overview of the EMF 27 study on global technology and climate policy strategies." *Climatic Change*.

Ley 24/2013 (2013). Ley 24/2013, de 26 de diciembre, del Sector Eléctrico. Boletín Oficial del Estado (BOE), Jefatura del Estado. Gobierno de España.

Ley 54/1997 (1997). Ley 54/1997, de 27 de noviembre, del Sector Eléctrico. *Boletín Oficial del Estado* (BOE), Jefatura del Estado. Gobierno de España.

Lilliestam, J. (2013). *Energy security and renewable electricity imports: the case of a Supergrid connecting Europe, North Africa and the Middle East.* Budapest, Department of Environmental Sciences and Policy, Central European University.

Lovins, A., I. Sheikh and A. Markevich (2008). *Nuclear power: Climate fix or folly?* Boulder, Rocky Mountain Institute.

Mau, P., J. Eyzaguirre, M. Jaccard, C. Collins-Dodd and K. Tiedemann (2008). "The 'neighbor effect': Simulating dynamics in consumer preferences for new vehicle technologies." *Ecological Economics 68*: 504-516.

McCollum, D. L., V. Krey and K. Riahi (2011). "An integrated approach to energy sustainability." *Nature Climate Change 1*: 428-429.

Messner, S. and L. Schrattenholzer (2000). "MESSAGE-MACRO: linking an energy supply model with a macroeconomic module and solving it iteratively." *Energy 25*: 267-282.

Messner, S. and M. Strubegger (1995). *User's Guide for MESSAGE III.* Laxenburg, Austria, International Institute for Applied Systems Analysis (IIASA)*:* 155.

MinEZ (2011). *Groene Groei. Ijsselstein, Ministerie van Economische zaken Landbouw en Innovatie:* 12.

MinEZ (2012a). *Overzicht van de openstelling SDE+ in fases.* Den Haag, Ministerie van Economische zaken Landbouw en Innovatie (MinEZ).

MinEZ (2012b). *Stimulering duurzame energieproductie.* Den Haag, Ministerie van Economische zaken Landbouw en Innovatie (MinEZ).

MinEZ (2013). *Stimulering duurzame energieproductie.* Den Haag, Ministerie van Economische zaken Landbouw en Innovatie (MinEZ).

MinEZ (2014). *Wetgevingsagenda STROOM.* Den Haag, Ministerie van Economische zaken Landbouw en Innovatie (MinEZ).

Munoz, L. A. H., J. C. C. M. Huijben, B. Verhees and G. P. J. Verbong (2014). "The power of grid parity: A discursive approach." *Technological Forecasting and Social Change.*

Nema, P., R. K. Nema and S. Rangnekar (2009). "A current and future state of art development of hybrid energy system using wind and PV-solar: A review." *Renewable and Sustainable Energy Reviews 13*: 2096-2103.

NREAP (2010). *Spain's National Renewable Energy Action Plan 2011-2020,* Ministerio de Industria, Turismo y Comercio. Gobierno de España.

Ondernemend Groen (2014). Zoekresultaat - Ondernemend Groen.

OSeMOSYS (2014). OSeMOSYS. *The Open Source Energy Modelling System* Retrieved 2014.04.08, from http://www.osemosys.org.

Pfenninger, S., A. Hawkes and J. Keirstead (2014). "Energy systems modeling for twenty-first century energy challenges." *Renewable and Sustainable Energy Reviews 33*: 74-86.

Prognos (2012). *Die Energieperspektivem für die Schweiz bis 2050: Energienachfrage und Elektrizitätsangebot in der Schweiz 2000-2050.* Bern, Prognos, Swiss Federal Office of Energy (BFE).

Rao, S., I. Keppo and K. Riahi (2006). "Importance of technological change and spillovers in long-term climate policy." *The Energy Journal 27*: 123-140.

RD (2004). Real Decreto 436/2004, de 12 *de marzo, por el que se establece la metodología para la actualización y sistematización del régimen jurídico y económico de la actividad de producción de energía eléctrica en régimen especial.* Boletín Oficial del Estado (BOE), Ministerio de Economía. Gobierno de España.

RD (2011). Real Decreto 1699/2011, de 18 de noviembre, por el que se regula la conexión a red de instalaciones de producción de energía eléctrica de pequeña potencia. *Boletín Oficial del Estado (BOE), Ministerio de Industria*, Turismo y Comercio. Gobierno de España.

RDL (2010a). Real Decreto 1565/2010, de 19 de noviembre, por el que se regulan y modifican determinados aspectos relativos a la actividad de producción de energía eléctrica en régimen especial. *Boletín Oficial del Estado* (BOE), Ministerio de Industria, Turismo y Comercio. Gobierno de España.

RDL (2010b). Real Decreto-ley 14/2010, de 23 de diciembre, por el que se establecen medidas urgentes para la corrección del déficit tarifario del sector eléctrico. *Boletín Oficial del Estado* (BOE), Jefatura del Estado. Gobierno de España.

RDL (2012). Real Decreto-ley 1/2012, de 27 de enero, por el que se procede a la suspensión de los procedimientos de preasignación de retribución y a la supresión de los incentivos económicos para nuevas instalaciones de producción de energía eléctrica a partir de cogeneración, fuentes de energía renovables y residuos. *Boletín Oficial del Estado* (BOE), Jefatura del Estado. Gobierno de España.

RDL (2013). Real Decreto-ley 9/2013, de 12 de julio, por el que se adoptan medidas urgentes para garantizar la estabilidad financiera del sistema eléctrico. *Boletín Oficial del Estado* (BOE), Jefatura del Estado. Gobierno de España.

REE (2013). *El Sistema Eléctrico Español. Avance del Informe*, Red Eléctrica de España, REE.

REP (2005). *Plan de Energías Renovables 2005-2010,* Instituto para la Diversificación y Ahorro de la Energía, IDAE.

REP (2011). *Plan de Energías Renovables 2011-2020,* Instituto para la Diversificación y Ahorro de la Energía, IDAE.

Rijksoverheid (2014). *Meer duurzame energie in de toekomst.*

Scheer, H. (2010). *Der energethische Imperativ.* 100% jetzt: wie der vollständige Wechsel zu erneuerbaren Energien zu realisieren ist. München, Kunstmann.

Schlesinger, M., P. Hofer, A. Kemmler, A. Kirchner, S. Strassburg, D. Lindenberger, M. Fürsch, S. Nagl, M. Paulus, J. Richter, J. Trüby, C. Lutz, O. Khorushun, U. Lehr and I. Thobe (2010). *Energieszenarien für ein Energiekonzept der Bundesregierung.* Basel, Köln, Osnabrück, Prognos AG, Energiewirtschaftliches Institut an der Universität zu Köln (EWI), Gesellschaft für wirtschaftliche Strukturforschung (GWS).

Scholz, Y. (2012). *Renewable energy based electricty supply at low costs.* Development of the REMix model and application for Europe. Stuttgart, University of Stuttgart.

SER (2013). *Energieakkoord voor duurzame groei.* Den Haag, Sociaal-Economische Raad (SER)*:* 146.

Swiss Federal Council (2013a). *Botschaft zum ersten Massnahmenpaket der Energiestrategie 2050 und zur Volksinitiative «Für den geordneten Ausstieg aus der Atomenergie».* Bern, Swiss Federal Council.

Swiss Federal Council (2013b). *Strategie Stromnetze; Detailkonzept im Rahmen der Energiestrategie 2050. Bern,* Swiss Federal Council.

Swiss Federal Parliament (2014). *Energiegesetz vom 26.* Juni 1998, version vom 1. Januar 2014 Bern, Swiss Federal Parliament.

Swissgrid (2014). *Kostendeckende Einspeisevergütung.* Retrieved 2014.03.31, from http://www.swissgrid.ch/swissgrid/de/home/experts/topics/renewable_energies/crf.html.

Thompson, M. (1984). "Among the energy tribes: a cultural framework for the analysis and design of energy policy." *Policy Sciences 17*: 321-339.

Trieb, F., T. Fichter and M. Moser (2014). "Concentrating solar power in a sustainable future electricity mix." *Sustainability Science 9*(1): 47-60.

Tweede Kamer (2001). *Uitvoeringsregeling energie-investeringsaftrek.* Den Haag, Tweede Kamer der Staaten-General.

Tweede Kamer (2007). *Besluit stimulering duurzame energieproductie.* Den Haag, Tweede Kamer der Staten Generaal.

Tweede Kamer (2009). *Subsidieregeling energie en innovatie.* Den Haag, Tweede Kamer der Staten Generaal.

Tweede Kamer (2012a). *Regeling openstelling en subsidieplafonds EZ 2012.* Den Haag, Tweede Kamer der Staaten-General.

Tweede Kamer (2012b). *Regeling openstelling en subsidieplafonds EZ 2013.* Den Haag, Tweede Kamer der Staaten-General.

Tweede Kamer (2013). *Tijdelijke regeling openstelling en subsidieplafonds EZ 2014.* Den Haag, Tweede Kamer der Staaten-General.

UNEF (2013). *Informe Anual UNEF 2013,* Unión Española Fotovoltaica (UNEF).

van der Sluijs, J., J. Potting, J. Risbey, D. van Vuuren, B. de Vries, A. Beusen, P. Heuberger, S. Corral Quintana, S. Funtowicz, P. Kloprogge, D. Nuijten, A. Petersen and J. Ravetz (2002). *Uncertainty assessment of the IMAGE/TIMER B1 CO2 emissions scenario, using the NUSAP method.* Utrecht, Department of science technology and society, Utrecht University.

van Vuuren, D. P., M. Isaac, M. G. J. den Elzen, E. Stehfest and J. van Vliet (2010). "Low stabilization scenarios and implications for major world regions from an integrated assessment perspective." *Energy journal 31*: 165-192.

Winskel, M., N. Markusson, H. Jeffrey, C. Candelise, G. Dutton, P. Howarth, S. Jablonski, C. Kalyvas and D. Ward (2014). "Learning pathways for energy supply technologies: Bridging between innovation studies and learning rates." *Technological Forecasting and Social Change 81*: 96-114.

Wynne, B. (1984). "The institutional context of science, models, and policy: the IIASA energy study." *Policy Sciences 17*: 277-320.

Zheng, C. and D. M. Kammen (2014). "An innovation-focused roadmap for a sustainable global photovoltaic industry." *Energy Policy 67*: 159-169.

In: Solar Power
Editor: Stephen Bailey
ISBN: 978-1-63321-317-3

Chapter 3

FABRICATION, ELECTRONIC PROPERTIES AND MICROSTRUCTURES OF TiO_2-BASED DYE-SENSITIZED SOLAR CELLS

Takeo Oku**, Nariaki Kakuta, Kengo Kobayashi and Atsushi Suzuki
Department of Materials Science,
The University of Shiga Prefecture, Shiga, Japan

ABSTRACT

TiO_2-based dye-sensitized solar cells were fabricated and characterized Solar cells mixed with xylenol orange and rose Bengal showed higher conversion efficiency compared to solar cells with a single dye. In addition, amorphous TiO_2 layers were introduced, which improved the conversion efficiency. Microstructures of the solar cells were examined by X-ray diffraction and transmission electron microscopy, and energy levels of the present solar cells were also discussed.

* Corresponding author: oku@mat.usp.ac.jp.

1. Introduction

Electrical energy storage systems have been developed, and solar cells are one of the good candidate clean energy devices because of no production of carbon dioxide causing global warming [1-3]. Although silicon (Si) solar cells have high conversion efficiency and a long lifetime, production processes are complicated and expensive.

Dye-sensitized solar cells, based on the concept of photo-sensitization of wide bandgap mesoporous oxide semiconductors [4], are now in a state of advanced development. This technology has been established as a promising low-cost photovoltaic concept [5], and has featured applications that can be colored and are lightweight. However, dye-sensitized solar cells have a short lifetime due to leakage and vaporization of the electrolytes. Therefore, studies of solidification of dye-sensitized solar cells have been performed [6, 7], and organic dyes without noble metals are expected as low-cost dyes.

The purpose of the present work is to investigate electrical and optical properties of dye-sensitized solar cells (DSSC), with an amorphous TiO_2 layer to introduce electrons at trap levels in acceptor and donor levels. Effects of organic dyes addition such as protoporphyrin IX (PPIX), xylenol orange (XO) and rose Bengal (RB) to dye-sensitized solar cells were fabricated and characterized.

2. Experimental Procedures

Schematic illustrations of the present solar cells are shown in Figure 1. Nanocrystalline TiO_2 photoelectrodes were prepared using the following procedure. TiO_2 powder was dispersed in an aqueous solution (1 ml) in a mixture of acetylacetone (0.02 ml, Shigma Aldrich) with Triton X-100 (0.01 ml, ICN Biomedicals) with polyethylene glycol (0.2 g, PEG#20,000, Nacalai Tesque) [8-10]. Atomic structure models of TiO_2 are shown in Figure 2. The TiO_2 paste was coated on pre-cleaned FTO by the squeegee method. After the TiO_2 coating, the FTO substrate was sintered for 30 minutes at 450 °C, and the prepared titanium isopropoxide (TTIP) solution dropped on the substrate. After that, the substrate was sintered for 60 min at 180°C. The TTIP solution was prepared by mixing TTIP (0.3 ml, Sigma Aldrich), acetilaceton (0.08 ml), ethanol (0.64 ml) and PEG (#2,000, 0.2 g, Nacalai Tesque). The TiO_2 electrodes were dissolved into the solved organic dyes such as, protoporphyrin

(Aldrich), xylenol orange and rose bengal solutions (Tokyo Chemical Industry) in distilled water, methanol or ethanol (Nacalai Tesque) [11-13]. Bellfine (0.1 g, Air Water) and Denka black (0.02 g, Denki kagakukogyo) as carbon was dispersed in the distilled water (0.8 ml) and ethanol (0.4 ml) with sodium carboxymethyl cellulose (0.012 g, Tokyo Chemical Industry) [14]. The carbon paste was applied on indium tin oxide (ITO, Geomatec) as an opposite electrode by the squeegee method.

The heat treatment of the carbon on the ITO substrate was carried out at 180 °C for 30 min. The electrolyte fabricated in a mixture of iodine (I_2, 0.05 g, Wako), lithium iodide (LiI, 0.09 g, Wako), ethylene carbonate (0.41 g, Shigma Aldrich Inc.), propylene carbonate (0.5 ml, Shigma Aldrich) and polyacrylonitrile (0.17 g, Shigma Aldrich) was agitated and heated at 80°C [15]. The dye-sensitized solar cells were assembled by putting the electrolyte between the adsorbed dye materials at the TiO_2 layer on the FTO glass substrate and the carbon film on the ITO substrate. Current density-voltage (J-V) characteristics (Hokuto Denko, HSV-100) of the solar cells were measured both in the dark and under illumination at 100mWcm^{-2} by using an AM 1.5 solar simulator (San-ei Electric, XES-301S) in N_2 atmosphere. The solar cells were illuminated through the side of the FTO substrates, and the illuminated area was 0.16 cm^2. Optical absorption of the solar cells was investigated by means of UV-visible spectroscopy (Hitachi, U-4100). Microstructures of the DSSC thin films were investigated by wide X-ray diffractometer (XRD, Philips, X'Pert-MPD System) with CuKα radiation operating at 40kV and 40mA, and transmission electron microscopy (TEM, Hitachi, H-8100). High-resolution TEM observation operated at 200 kV was also performed, which could provide structural information in nanoscopic scale [16, 17].

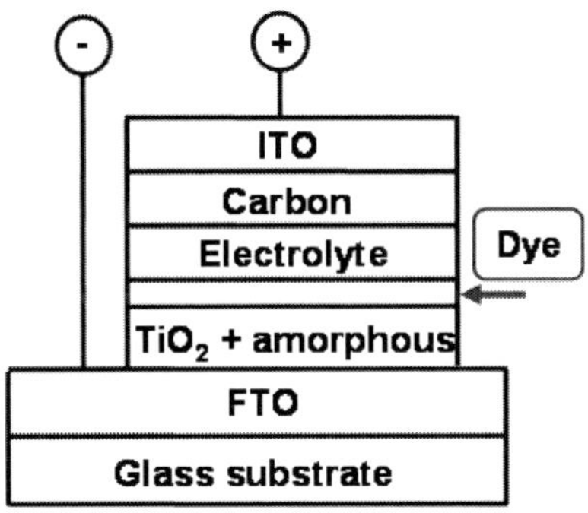

Figure 1. Structures of dye-sensitized solar cells with dyes.

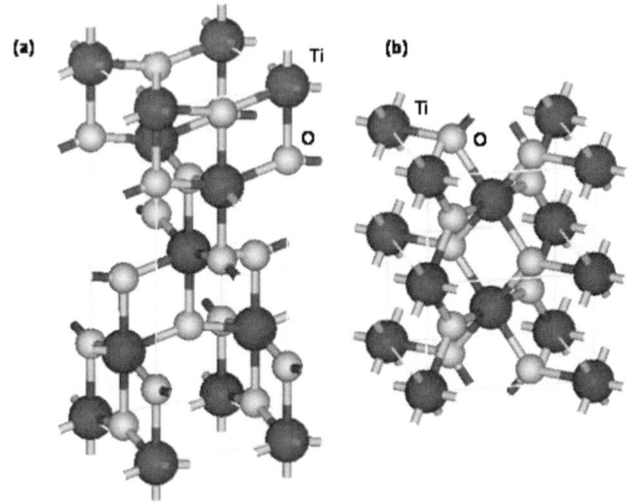

Figure 2. Structure models of TiO2 with (a) anatase and (b) rutile structures.

3. Results and Discussion

Measured cell performance of DSSCs is summarized as Table 1. Experimental parameters of open circuit voltage (Voc), short circuit current (Jsc), fill factor (FF) and conversion efficiency are listed in Table 1. A normal DSSC with PPIX provided an efficiency of 2.3×10^{-3}.

Table 1. Measured parameters of DSSCs

DSSC samples	V_{OC} (V)	J_{SC} ($mAcm^{-2}$)	FF	η (%)
PPXI	0.091	0.088	0.29	2.3×10^{-3}
XO + RB without amorphous TiO_2	0.36	0.60	0.56	0.12
XO + RB with amorphous TiO_2 layer	0.33	0.83	0.58	0.16

A TEM image and an electron diffraction pattern of the DSSC with PPIX are shown in Figure 3(a) and 3(b), respectively. The TEM image indicates TiO_2 naonparticles with sizes of 20 nm ~60 nm, and the electron diffraction pattern shows 101, 103 and 200 reflections of a tetragonal TiO_2 anataze phase. A high-resolution electron microscopy (HREM) image of an interface of TiO_2 nanoparticles is shown in Figure 3(c). The interface is directly connected in atomic scale, which would results in good carrier transport between TiO_2

nanoparticles. A HREM image of surface of the TiO_2 nanoparticle is shown in Figure 3(d), and the TiO_2 nanoparticle is covered by a ~2 nm thick PPIX layer with an amorphous structure.

Figure 4 is an energy level diagram of the present DSSC with PPIX. Voc of the present DSSC would be related with the energy gap (E_g) between HOMO of PPIX and conduction band of TiO_2. Control of the energy gap is important for improving the efficiency. The light irradiation caused charge-separation at PPIX, and the electrons are transferred to conductive band of TiO_2. Holes at HOMO level of PPIX lead to conversion of I^- to I^{3-} in the electrolyte, and I^- reduced the oxidized porphyrin to porphyrin with normal valence. Measured J-V characteristics of a DSSC with an amorphous TiO_2 layer are shown in Figure 5. The current density of DSSC with an amorphous TiO_2 layer was higher than that of DSSC without an amorphous TiO_2 layer. The cell performance is summarized as Table 1. The current density was improved from 0.6 $mAcm^{-2}$ to 0.83 $mAcm^{-2}$. The other parameters are almost the same. As a result, the conversion efficiency was improved from 0.12 % to 0.16 %.

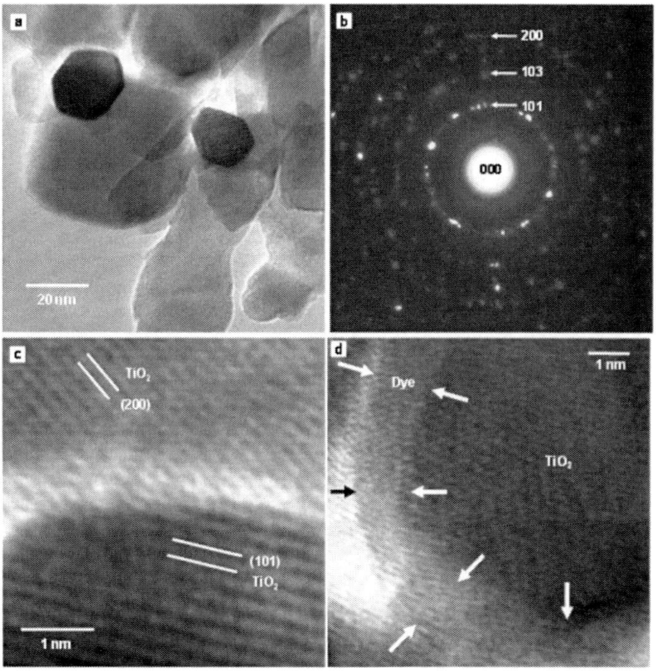

Figure 3. (a) TEM image and (b) electron diffraction pattern of TiO_2 DSSC with protoporphyrin. HREM images of (c) interface and (d) surface of TiO_2 nanoparticles.

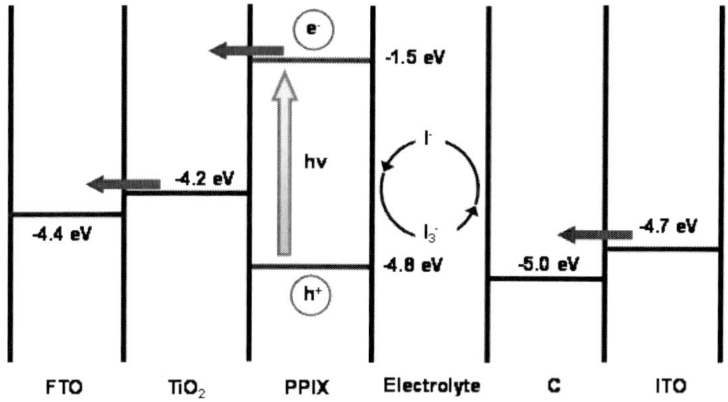

Figure 4. Energy level diagram of the present DSSC with PPIX.

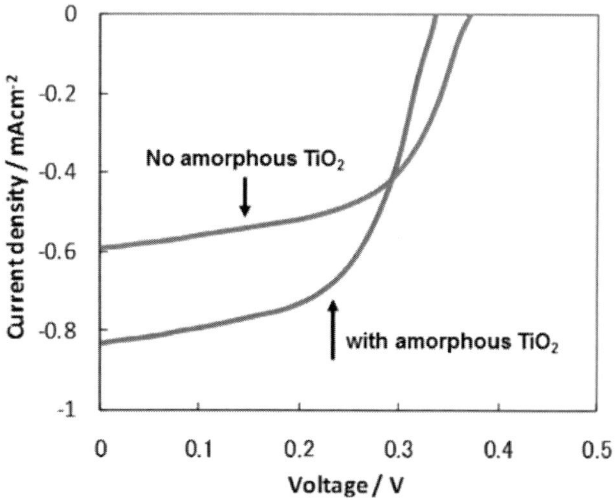

Figure 5. J-V characteristics of DSSCs with or without an amorphous TiO_2 layer.

Figure 6 shows measured optical absorption of DSSC with an amorphous TiO_2 layer compared to that without an amorphous TiO_2 layer. An optical absorption peak of TiO_2 around 360 nm is increased by introducing the amorphous TiO_2 layer.

Figure 7 shows X-ray diffraction patterns of TiO_2 layers prepared from the TTIP solution as a function of temperature. No peak of an anatasephase was observed when annealing at 250°C. Small peaks of the anatase phase were observed after annealing at 350°C. Diffraction peaks of TiO_2 (P25) are shown for comparison with the amorphous TiO_2.

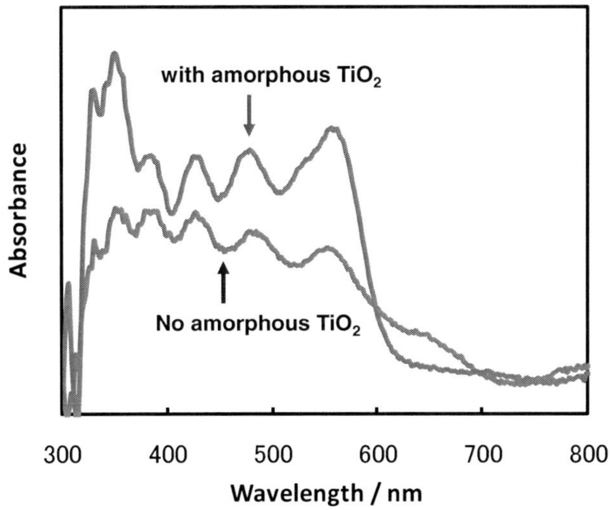

Figure 6. Optical absorption of DSSCs with or without an amorphous TiO_2 layer.

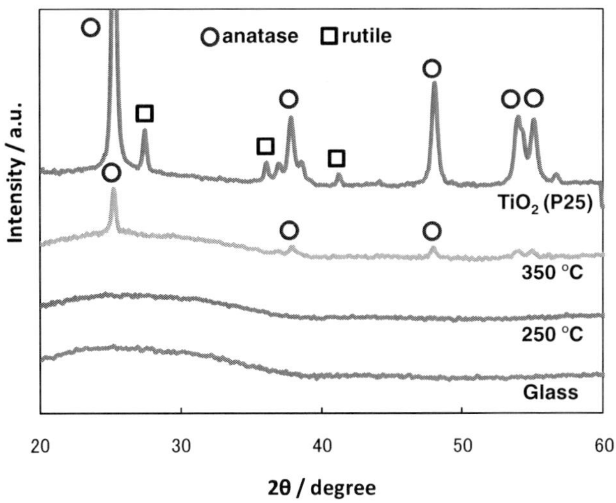

Figure 7. X-ray diffraction patterns of TiO_2 layers as a function of temperatures.

Figure 8 shows a TEM image and an electron diffraction pattern of the amorphous TiO_2 layer. The amorphous TiO_2 layer around the TiO_2 nanoparticle is observed in the TEM image. As shown in Figure 8(b), the electron diffraction pattern indicates 101 and 103 reflections of the polycrystalline coagulation of TiO_2 with a tetragonal anatase structure. Diffuse ring of an amorphous phase was also observed, which indicates the structure of the TiO_2 layer was a mixture of anatase and amorphous phases.

An energy level diagram of DSSC with the amorphous TiO_2 layer is summarized in Figure 9. The values of HOMO and LUMO were calculated by Gaussian 03 (B3LYP, 6-31G*). An energy barrier would exist at the semiconductor metal interface.

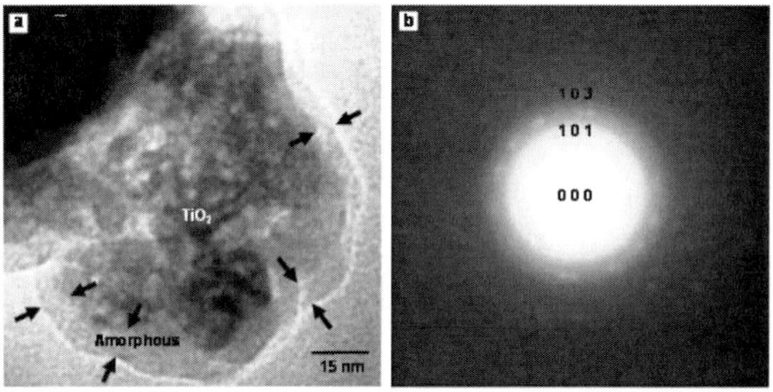

Figure 8. (a) TEM image and (b) electron diffraction pattern of an amorphous TiO_2 layer.

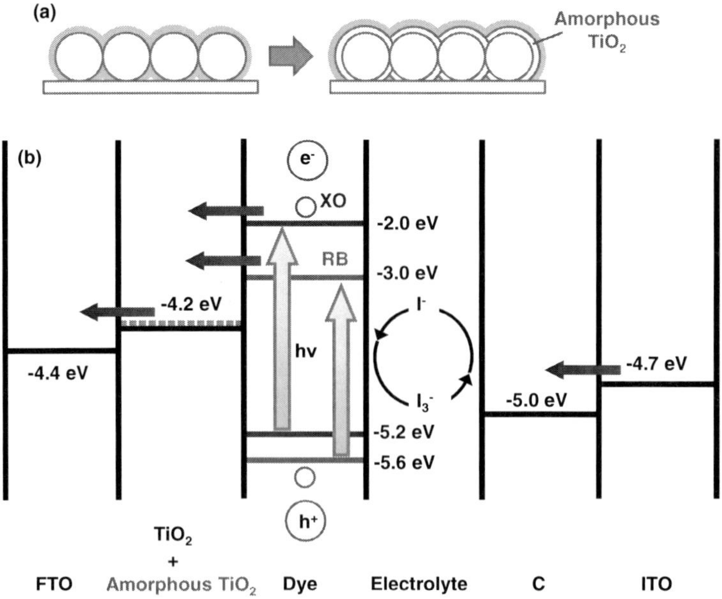

Figure 9. (a) Schematic diagram of TiO_2 electrode with an amorphous TiO_2 layer. (b) Energy level diagram of the present DSSC with an amorphous TiO_2 layer.

Electronic charge generation was caused by light irradiation from the FTO substrate side. The TiO_2 layer or amorphous TiO_2 layer receives the electrons from the dye, and the electrons are trapped by several trap levels of the amorphous TiO_2. Electrons are transported to an FTO electrode through the TiO_2 layer, and electrons are transferred to the outside circuit, and flow through the carbon electrode. Then, electrons return to the electrolyte by oxidation-reduction reaction. The improvement of the present solar cells is due to the introduction of the amorphous TiO_2 electrode, and this structure should be investigated further.

Carrier recombination would be a main cause for low conversion efficiency. The improvement of the conversion efficiency can be expected by improving the charge separation and the electronic charge transfer.

Improvements of the conversion efficiency by coating materials on TiO_2 have been reported. Insulator oxides such as Al_2O_3 or SiO_2 on TiO_2 have been introduced forDSSCs [18]. These oxides suppress carrier recombination by reverse transfer of electrons in the cells. In the present study, DSSCs with amorphous TiO_2 layers on TiO_2 were fabricated, and the amorphous TiO_2 would attract electrons because of many trap levels in the acceptor and donor levels. In addition, effectiveness of carrier separation is high because of large contact area. Further improvement would be possible by introducing nanoparticles as light harvesting materials [19, 20].

CONCLUSION

Amorphous TiO_2 was introduced into DSSC with mixed xylenol orange and rose Bengal. The optical absorption peak of TiO_2 around 360 nm was increased by introducing the amorphous TiO_2 layer. Diffraction spots of the anatase phase and a diffuse ring from the amorphous phase were observed for the TEM observations.

The proposed energy diagram shows the amorphous TiO_2 would attract electrons at several trap levels in the TiO_2 layers. In addition, the carrier separation is high, which would be due to large contact area. As a result, the current density was improved from 0.60 $mAcm^{-2}$ to 0.83 $mAcm^{-2}$, and the conversion efficiency was improved from 0.12 % to 0.16 % by the introduction of the amorphous TiO_2 layer.

References

[1] Green, M. A., Emery, K., King, D. L., Hishikawa, Y.,Warta, W. (2006). *Prog. Photovolt. Res. Appl.*, *14*, 455.

[2] Oku, T., Takeda, A., Nagata, A., Noma, T., Suzuki, A. and Kikuchi, K. (2010). *Energies*, *3*, 671.

[3] Oku, T., Takeda, A., Nagata, A., Kidowaki, H., Kumada, K., Fujimoto, K., Suzuki, A., Akiyama, T., Yamasaki, Y., Ōsawa, E. (2013). *Mater. Technol.*, *28*, 21.

[4] Chen, H., Cong, T. N., Yang, W., Tan, C., Li, Y., Ding, Y. (2009). *Prog. Nat. Sci.*, *19*, 291.

[5] Durr, M., Bamedi, A., Yasuda, A., Nellesa, G., (2004). *Appl. Phys. Lett.*, *84*, 3397.

[6] Murai, S., Mikoshiba, S., Hayase, S. (2007). *Sol. Energy Mater. Sol. Cells*, *91*, 1707.

[7] Schmidt-Mende, L., Bach, U., Humphry-Baker, R., Horiuchi, T., Miura, H., Ito, S., Uchida, S., Grätzel, M., (2005). *Adv. Mater. 17*, 813.

[8] Lewis, L. N., Spivack, J. L., Gasaway, S., Williams, E. D., Gui, J. Y., Manivannan, V.,Siclovan, O. P.(2006). *Sol. Energy Mater. Sol. Cells*, *90*, 1041.

[9] Nazeeruddin, M.K., Kay, A., Rodicio, I., Humphery-Baker, R., Muller, E., Liska, P., Vlachopoulos, N., Grätzel, M. (1993). *J. Am. Chem. Soc.*, *115*, 6382.

[10] Oku,T., Kakuta, N., Kobayashi,K., Suzuki, A., Kikuchi,K. (2011). *Prog. Nat. Sci.*, *21*, 122.

[11] Pradhana, B., Batabyala, S. K., Pal, A. J. (2007). *Sol. Energy Mater. Sol. Cells*, *91*, 769.

[12] Kakuta, N., Oku, T., Suzuki, A., Kikuchi, K., Kikuchi S. (2009). *J. Ceram. Soc. Jpn. 117*, 964.

[13] Matsubara, T., Ichikawa, Y., Aramaki, K., Katagiri, A. (2005). *Sol. Energy Mater. Sol. Cells*, *85*, 269.

[14] Imoto, K., Takahashi, K., Yamaguchi, T., Komura, T., Nakamura, J., Murata, K., (2003). *Sol. Energy Mater. Sol. Cells*, 79, 459.

[15] Ileperuma, O.A., Dissanayake, M.A.K.L., Somasunderam, S., Bandara, L.R.A.K. (2004). *Sol. Energy Mater. Sol. Cells*, *84*, 117.

[16] Oku, T. (2011). *Nanosci. Nanotech. Asia, 1*, 59.

[17] Oku, T. (2012). *Nanotechnol. Rev.,1*, 389.

[18] Palomares, E., Clifford, J. N., Haque, S.A., Lutz, T., Durrant, J.R. (2003). *J. Am. Chem. Soc.*, *125*, 475.

[19] Nagata, A., Oku, T., Kikuchi, K., Suzuki, A., Yamasaki, Y., Ōsawa, E. (2010). *Prog. Nat. Sci.*, *20*, 38.

[20] Park, N. G., Kang, M. G., Ryu, K. S., Kim, K. M.,Chang, S. H. (2004). *J. Photo-Chem. Photobiol. A*, *161*, 105.

In: Solar Power
Editor: Stephen Bailey

ISBN: 978-1-63321-317-3

Chapter 4

SOLAR TOWER POWER PLANTS PERFORMANCE AND RELIABILITY ANALYSIS

S. Benammar[1,*], A. Khellaf[2,†] and K. Mohammedi[1]

[1]Laboratoire Energétique Mécanique et Ingénieries (LEMI),
M'hamed Bougara University, Boumerdes, Algeria
[2]Centre de Développement des Energies Renouvelables
CDER Route de l'Observatoire Bouzaréah, Algeria

ABSTRACT

The objective of the present study is to address the main methods and tools used to analyze the reliability and the performance of solar tower power plants (STPP).

An overview of STPP technology, current status and a presentation of the major components including the heliostat field, the solar receiver and the power conversion system are first highlighted. Next, the analysis of the performances and the reliability of STPP are outlined. Artificial neural network method (ANN) is used to assess the perfor-mance of STPP. In this study, the STPP receiver efficiency is considered as a performan-ce metric. In order to show the efficacy of this method, an example of hypothetical STPP is proposed. After the investigation of the simulation result, a very good learning and accuracy of ANN are noted.

[*] S. Benammar: Laboratoire Energétique Mécanique et Ingénieries (LEMI), M'hamed Bougara University - Boumerdes, 35000 Boumerdes, Algeria. E-mail: sa.benammar@gmail.com.

[†] A. Khellaf: Centre de Développement des Energies Renouvelables CDER Route de l'Observatoire Bouzaréah 16340 Algiers, Algeria. E-mail: khellaf.at@gmail.com.

The reliability analysis methods are also investigated. The different models and tools, such as sofwares, for STPP reliability and performance evaluation are presented. The study will focus on specific methods such as the First Order Reliability Method (FORM) for the reliability and performance analysis of STPP. Detailed analysis concerning FORM is developed in this chapter. The hypothetical STPP example is also proposed to evaluate this method. The results show that FORM is suitable to assess the reliability of STPP systems and it converges quickly. It can also be used to adjust the first design parameters to a new safe design.

1. INTRODUCTION

Dominated by coal and hydrocarbon energy sources, the power generation sector is actually facing many challenges. Indeed as the demand is steadily increasing, concern about the depleting energy sources is increasing. Moreover, the use of the conventional energy sources is responsible for the environment degradation as a result of pollutants emission. To then effectively address these challenges, steps have to be taken to introduce cleaner and sustainable energy sources. As a result, interest in renewable energy and more particularly solar thermal power has been growing.

In solar thermal power, the thermal energy necessary for electricity generation is produced by concentrating and transforming into heat the solar direct normal irradiation. There are different concentrating solar power technologies. The most advanced technologies are those of the parabolic trough, the solar tower, the parabolic dish and the linear Fresnel reflector. The use of solar tower power plants (STPP) for electricity production promises to be one of the most economic options to replace conventional power plants. Indeed, research and development activities on its basic subsystems have been increasing.

Modeling the performance and the economics of a solar power plants based on technolo-gies such as power towers and parabolic troughs has evolved over several decades.

Full performance models for parabolic trough plants with no thermal energy storage (TES) system have been presented by Lippke (Lippke, 1995), Price et al. (Price et al., 1995), Patnode (Patnode, 2006) and Jones et al. (Jones et al., 2001).

Price (Price, 2003) has expanded his early work (Price et al., 1995) by including a TES system in the simulation. A detailed performance model for a parabolic trough plant with TES together with a comparison of the model results to measured data from an operating plant of this type has been developed (Isabel et al., 2011).

There are several documented complete models for trough plants with storage, namely:

- SOLERGY that has been developed by Sandia National Laboratories in the US. It was originally a model for tower plants but has been updated over the years to accurately model trough plants (Stoddard et al., 1987);
- System Advisor Model (SAM) that has been developed by the National Renewable Energy Laboratories in the US (Price, 2003; Blair et al., 2008a; Blair et al., 2008b; Wagner et al., 2010);
- Greenius that has been developed by the German Aerospace Centre (Dersch et al., 2008; Hennecke et al., 2010).

A method (Montes et al., 2009) has been developed to assess the influence of the solar multiple on the annual performance of parabolic trough solar thermal power plants with direct steam generation (DSG). This method of assessing the annual plant performance, using only one clear day and one cloudy day in every month can be applied for sites where only limited meteorological data are available. Economic analysis has been carried out; the levelized cost of electricity (LEC) has been determined for all the solar multiples considered. It has been observed that the levelized cost of electricity gets higher as the solar multiple of the power plant increases, mainly owing to the great investment cost in the solar field.

Using advanced software tools, the design and performance assessment has been carried out for several prototype solar tower gas turbine power plants in the power levels of 1 MW, 5MW and 15MW (Peter et al., 2006). The LEC has then been evaluated. It has been found that the larger units (power larger than 10 to 15 MW), especially combined cycle systems, show very low cost for solar produced bulk electricity. The cost decreases with increasing power level. Small-scale units (power smaller than 5 to 10 MW) should be applied in distributed markets using cogeneration to start market introduction.

The probabilistic performance model of concentrating solar power plants including quantification of uncertainties inherent in the system has also been considered (Ho and Kolb, 2010). Two examples, a simple cost model and a more detailed performance model of a hypothetical 100-MWe power tower, are provided to illustrate the methods. The LEC equation, including annualized capital costs, annual operation and maintenance costs, annual energy generated and system availability, has been calculated 300 times using randomly sampled values from the distribution of input parameters to yield a distribution of equally probable LEC values. The distribution of calculated LEC values ranged from approximately 0.08 $/kWh to 0.16 $/kWh.

The thermodynamic performance of the Kalina cycle has been evaluated in terms of the plant exergy efficiency for a central receiver solar thermal power plant with direct steam generation (Modi and Haglind, 2014). The result has been compared to a simple Rankine cycle. The rates of exergy destruction for the different components in the two cycles were also calculated and compared. The results suggest that the simple Rankine cycle exhibits better performance than the Kalina cycle when the heat input is only from the solar receiver. However, when using a two-tank molten-salt storage system as the primary source of heat input, the Kalina cycle showed an advantage over the simple Rankine cycle because of about 33 % reduction in the storage requirement.

A mathematical model based on energy analysis for modeling and simulating of STPP performances without energy storage has been proposed (Benammar et al., 2014). The analysis of the obtained results shows the existence of an optimal receiver efficiency value for each steam mass flow, receiver surface temperature and receiver surface area.

The cost and performances of a solar tower power plant with integrated latent thermal energy storage have been analyzed (Nithyanandam and Pitchumani, 2014). As performance metrics, the capacity factor, the LCE (levelized cost of electricity), the storage capital cost and the annual exergetic efficiency have been used. The effects of various storage system design parameters on these performance metrics have also been investigated.

This chapter describes the main methods and tools used to analyze the performance and reliability of STPP. It begins by representing the STPP system, their functioning mode and their components. Then, STPP performance and economic analysis is presented. To this end, the computer codes used to estimate the annual energy and costs are described. Besides, a simulation model based on artificial neural networks (ANN) method that reproduces the performance of STPP is detailed.

The aim of this model is to facilitate the prediction of the efficiency of these plants. The ANN data are given from mathematical models developed by Benammar et al. (Benammar et al., 2014) using energy analysis. After that, STPP reliability analysis is developed in this chapter in order to evaluate the availability and failure probability of the system. In this study, Pro-Opta

Reliability Analysis Software and first order reliability method (FORM) have been treated.

2. Solar Tower Power Plant Description

As shown in Figure 1, solar tower power plants use hundreds to thousands of large sun-tracking flat heliostats (mirrors) to concentrate sunlight onto a receiver placed on top of a tall tower. The tower height would be proportional to the energy production of the plant. Computer controlled heliostats track the sun and reflect the sunlight to the receiver. The complete group of heliostats is called the collector field. Depending on the configuration of the receiver, there are two main collector field lay outs. The collector field may surround the tower or it may be located only on one side of the tower. For a one-side lay out, the field lies north of the tower for a site located in the northern earth hemisphere; it lies south of the tower for a site located in the southern earth hemisphere. A power tower system needs preferably relatively flat terrain, with a slope of one to two percent at most.

In the receiver, the collected solar radiation is converted to heat in a receiver fluid such as water/steam, liquid sodium, or molten nitrate salt flowing through small receiver tubes. If water/steam is the receiver fluid, the steam may be sent directly to the turbine generator. If one of the other receiver fluids is used, the energy in the fluid must be transferred to water/ steam by means of heat exchangers before being used to generate electricity in the turbine generator.

Molten salt is the most used fluid not only as heat transfer fluid but also as storage fluid. Power tower systems would use molten salt primarily because of its superior heat-transfer and energy-storage capabilities. Molten salt is typically a mixture of 60% sodium nitrate and 40% potassium nitrate. Molten salt as a mixture of 48% calcium nitrate, 7% sodium nitrate, and 45% potassium nitrate is also being investigated. The molten salt can be stored at tempera-tures of 565 °C in a storage tank for use on cloudy days or after the sun has set.

Also, a power tower, using molten salt rather than another type of HTF, can operate at a much higher temperature range which can improve the efficiency of the steam turbine genera-tor (e.g., from 37.6% to 40%). The marginal cost of collecting and storing this energy is less than the cost of increasing turbine size to match the peak thermal output.

Determination of the optimum storage size to fulfill the energy dispatch requirements of a particular application is a part of the central receiver design process (Falcone, 1986).

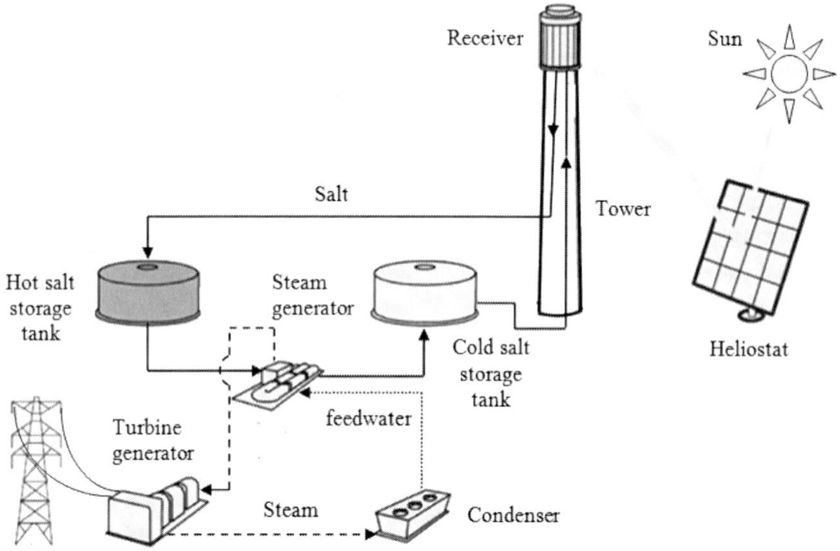

Figure 1. Schematic of hypothetical molten-salt central receiver system with thermal storage.

3. Performance Analysis

The performance of any power plant can be determined by four factors:

(a) Capability to satisfy functional needs;
(b) Efficiency to effectively utilize the energy supplied;
(c) Reliability to start or continue to operate;
(d) Maintainability to quickly return to service after one failure.

It is evident that the first two factors are influenced by the power plant design, manu-facturing, construction, operation and maintenance. Capability and efficiency reflect how well the power plant is designed and constructed. On the other hand, the last two factors, reliability and maintainability, are operations-related issues and are influenced by the power plant potential to remain operational and the efficiency of the repair aiming at returning to service process. It would be conceivable to have a power plant that is highly reliable, but does not achieve high performance.

Usually the power plant performance analysis is based on thermodynamic parameters evaluation in order to define the plant operational efficiency. Different methods have been used by scholars in order to evaluate the solar power plant efficiency. These methods are summarized into two main categories: analytical methods based on the physical models and artificial intelligence (AI) methods requiring the experimental data. Several softwares used to estimate the solar power plant efficiency, such as SOLARGY, TRNSYS, DELSOL and SAM, are based on the analytical methods. The most AI methods which have been used to estimate and predict the performance of solar power plants are the artificial neural networks (ANN).

In this part of chapter we have treated in details only the ANN method with an overview of some of the mostly used softwares (SOLARGY, TRNSYS, DELSOL and SAM).

3.1. Performance Indexes

3.1.1. Overall Equipment Effectiveness (OEE)

Process companies are adopting a new consolidated approach to performance improve-ment. This approach is based upon the use of a KPI (key performance indicator) known as Overall Equipment Effectiveness (OEE).

OEE can be calculated as:

$$\text{OEE}=\text{Availability}\times\text{performance}\times\text{Quality} \tag{1}$$

The performance of the thermal power plant can be represented by its efficiency. The efficiency of a power plant is usually measured as a ratio of its electrical output to the amount of heat used, expressed as a percentage. Typical commercial plants range is about 30–65% efficiency. The more efficient plants cost more to build. Efficiency depends more on how the energy is used rather than how it is produced because ratings are based on conversion of heat to electrical power.

The quality of the power plant is associated with the parameters (voltage and frequency) of the generated electricity in comparison with the required standards.

The availability of the power plant is associated with the reliability and maintenance planning of each piece of equipment installed in the plant. The availability depends also on the skills of operators and maintenance teams.

The use of OEE index can help the electricity-generating power-stations managers to investigate theirs competence in maintaining reliable equipment at competitive costs.

3.1.2. Levelized Energy Cost (LEC)

Levelized Energy Cost (LEC, also known as Levelised Cost of Energy, abbreviated as LCOE) is the price at which electricity must be generated from a specific source to break even over the lifetime of the project.

It is an economic assessment of the cost of the energy-generating system. It includes all the costs over its lifetime: initial investment, operations and maintenance, cost of fuel, cost of capital. It is very useful in calculating the costs of generation from different sources.

It can be defined in a single formula as:

$$LEC = \frac{\sum_{t=1}^{n} \frac{I_t + M_t + F_t}{(1+r)^t}}{\sum_{t=1}^{n} \frac{E_t}{(1+r)^t}} \qquad (2)$$

where

LEC: Average lifetime levelized electricity generation cost,
I_t: Investment expenditures in the year *t*,
M_t: Operations and maintenance expenditures in the year *t*,
F_t: Fuel expenditures in the year *t*,
E_t: Electricity generation in the year *t*,
r: Discount rate.

The LEC can also be estimated using the simplified equation proposed by Becker and Klimas (Becker and Klimas, 1993), and Ho and Kolb (Ho and Kolb, 2010) which is as follows:

$$LEC = \frac{ACC + AOMC}{AEG \times A} \tag{3}$$

ACC: Annualized capital cost,
AOMC: Annual operations and maintenance cost,
AEG: Annual energy generated,
A: Availability.

This formula has been updated to take into account the additional parameters for indirect charges and financing (Ho and Kolb, 2010). The following equation was suggested for the evaluation of the annualized capital costs reported in Equation (3):

$$ACC = FCR \times DC \times (1 + INDC) \times (1 + AFUDC) \tag{4}$$

where

FCR: Constant-dollar fixed charge rate,
DC: Total direct costs,
INDC: Indirect charges,
AFUDC: Allowed funds during construction to cover interest charges.

3.2. Solar Tower Power Plant Performances

To date, interest has been concentrated on the evaluation of the performance and on thermal modeling of the solar tower power plant. In general, the previous research studies can be grouped into two types. The first group focuses mainly on the thermal losses analysis of the solar cavity receiver and various analytical models for the estimation of thermal losses were proposed (Li et al., 2010; Lu and Ding, 2011; Zhang et al., 2013). The second group, on the other hand, has concentrated its solar plant modeling effort mainly on the modular modeling method of the system. In these studies, the main objective was to calculate the power output and the efficiency of the solar power plant using the energy balance (Chao et al., 2011; Ershu et al., 2011; Qiang et al., 2012; Benammar et al., 2014).

In the last decade, the use of artificial intelligence methods has been increasing gradually. ANN has been becoming increasingly popular in thermal engineering applications. Several studies have been introduced about using

ANN in thermal applications (Kalogirou, 2001; Kalogirou, 2006; Kurt et al., 2008; Esen et al., 2008a; Esen et al., 2008b; Caner et al., 2011; Benazzouz et al., 2011; Benammar and Benazzouz, 2012). In this section, a feed-forward neural network based on back propagation algorithm is developed to predict the thermo-dynamic performances of solar central cavity receiver. Multiple Layer Perceptron has been considered. The network has been trained using numerical results from mathematical models developed by Benammar et al. (Benammar et al., 2014).

3.2.1. Artificial Neural Networks (ANN)

A. ANN Analysis

Neural networks are composed of simple elements operating in parallel. These elements are inspired by biological nervous systems.

As in nature, the network function is determined largely by the connections between elements. A neural network can be trained to perform a particular function by adjusting the values of the connections (weights) between the elements.

Commonly, neural networks are adjusted, or trained, so that a particular input leads to a specific target output.

MLP network consists of an input layer of source neurons, one or more hidden layers and an output layer. The general formulas used to calculate the hidden layer and output layer neurons, are respectively (Benazzouz et al., 2011; Benammar and Benazzouz, 2012):

$$u_j = F_1\left(\sum_{i=1}^{M} w_{i,j}^1 x_i + b_j^1\right) \tag{5}$$

$$Y_k = F_2\left(\sum_{l=1}^{N} w_{l,j}^2 u_l + b_k^2\right) \tag{6}$$

F_1 and F_2 are the activation functions. Generally, a sigmoid activation function (equation (7)) is used in the hidden layer and a linear activation function is used (equation (8)) in the output layer. In this study case, the sigmoid activation function (equation (9)) has given best results than the linear activation function. $w^1_{i,j}$ and $w^2_{i,j}$ are the weight matrixes in the input and output layer respectively. b^1_j and b^2_k are the bias vectors in the input and output layer respectively. x_i is the input vector (data vector).

$$F_1(x) = \frac{2}{1+e^{-2x}} - 1$$

(7)

$$F_2(x) = x \tag{8}$$

$$F_1(x) = \frac{e^x}{(e^x+1)} = \frac{1}{1+e^{-x}} \tag{9}$$

MLP network learning involves finding the weight values $w^1_{i,j}$ and $w^2_{i,j}$ of the connecting neurons that minimize the mean square error function MSE between the actual product (estimated) and target values (desired) in the corresponding training set. For the network learning, we have used the retro-propagation algorithm. More details have been given in a previous work (Benammar and Benazzouz, 2012).

Mathematically, the training process is a nonlinear optimization problem which requires minimizing an objective (cost) function. This objective function (MSE) is an error measure between the predicted outputs Y and the target outputs T.

The mean square error (MSE) is estimated using the following relation:

$$MSE = \frac{1}{n}\sum_{k=1}^{n}\left(T_k - Y_k\right)^2 \tag{10}$$

where T_k is the kth output target or observed value, Y_k is the kth predicted value and n is the number of output data.

B. Designing ANN Models

After the preparation and the collection of input and output data, three data pre-processing procedures are conducted to train the ANNs more efficiently. These procedures are: (1) normalization of the data, (2) randomization of the data and (3) separation of the data set. Normalization procedure before presenting the input data to the network is generally a good practice, since mixing variables with large magnitudes and small magnitudes will confuse the learning algorithm on the importance of each variable and may force it to finally reject the variable with the smaller magnitude (Tymvios et al., 2008; Al-Shamisi et al., 2011). Randomization of the data is necessary to ensure the best learning and validation of network. The prepared data must be separated into test data (20%) and training data (80%).

During the training process, the weights are adjusted in order to make the actual outputs (predicated) close to the target (measured) outputs of the network. The next step is to test the performance of the developed model.

At this stage unseen data are exposed to the model. The flowchart presented in Figure 2 shows the main steps for the designing of ANN models.

In order to evaluate the performance of the developed ANN models quantitatively and check whether there is any underlying trend in performance of ANN models, statistical analysis involving the coefficient of determination (R^2), sum squared errors (SSE), the root mean square error (RMSE), and the mean relative errors (MRE) have been conducted.

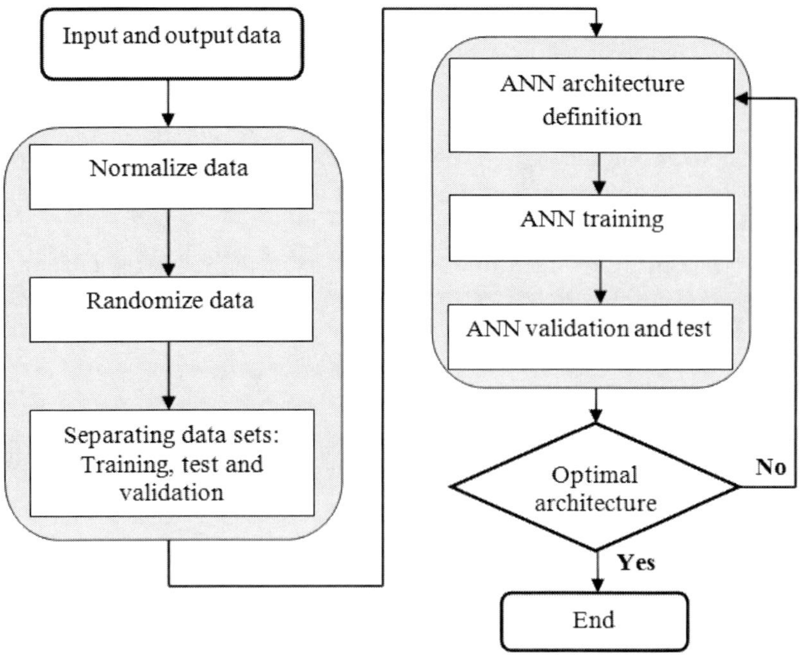

Figure 2. Flowchart with data pre-processing and artificial neural network training.

C. Definition of the Statistical Error Parameters

The prediction performances of the networks has been evaluated using the sum squared errors (SSE), the root mean square of the error (RMSE), the coefficient of multiple determina-tions (R^2) and the mean relative errors (MRE).

The sum squared errors (SSE) measures the performance according to the sum of the squared errors. The root mean square error (RMSE) (also called the root mean square deviation, RMSD) is a frequently used measure of the difference between values predicted by a model and the values actually observed from the system that is being modelled.

These individual differences are also called residuals, and the RMSE serves to aggregate them into a single measure of predictive power. SSE and RMSE are calculated via Equations (11) and (12).

$$SSE = \sum_{k=1}^{n} (T_k - Y_k)^2 \tag{11}$$

$$\%RMSE = \left(\frac{SSE}{n}\right)^{1/2} \times 100 \tag{12}$$

In addition, the coefficient of multiple determinations (R^2) is calculated according to Equation (13).

$$R^2 = 1 - \frac{SSE}{\sum_{k=1}^{n} p_k^2} \tag{13}$$

The mean relative errors (MRE) are computed with Equation (14).

$$MRE(\%) = \frac{1}{n}\sum_{k=1}^{n}\left(100 \times \frac{|T_k - Y_k|}{T_k}\right) \tag{14}$$

D. Example: Solar Tower Power Plant

A simple performance model of a hypothetical 100 kW power tower is provided to illustrate the method. Figure 3 represents a schematic of the hypothetical solar tower power plant considered in this example.

Table 1 and Table 2 show the main design and exploitation characteristics of this hypothetical STPP.

In this work the prediction ANN system contains three input vectors (receiver surface temperature, receiver surface area and total heliostat area) and the performance parameter (efficiency) has been selected as the output vector.

The input and output data are given from the numerical results of the mathematical model developed by the authors in a previous study (Benammar et al., 2014).

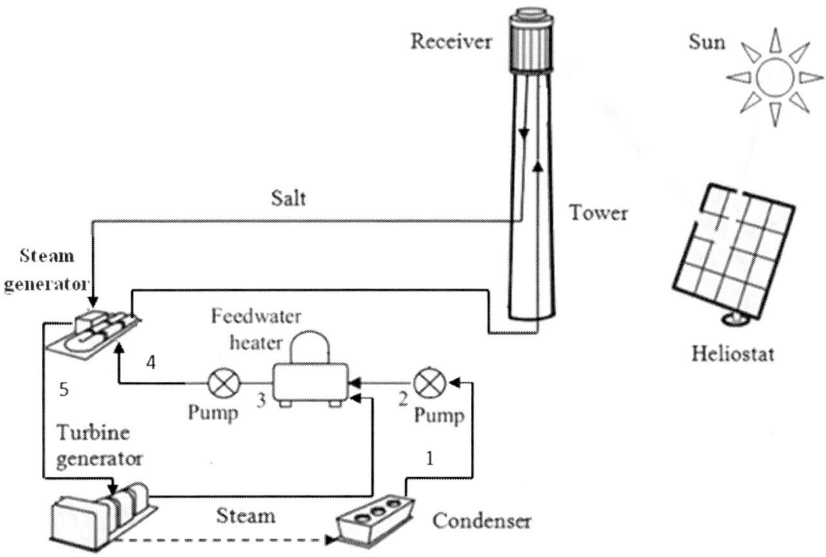

Figure 3. Schematic of hypothetical solar tower power plant.

Table 1. Main design characteristics of STPP

Parameters	Value	Unit
Tube diameter	0.019	m
Tube thickness	0.00165	m
Tube conductivity	23.9	W/m.K
Emissivity	0.8	-

Parameters	Value	Unit
Reflectivity	0.04	-
Insulation layer thickness	0.07	m
Aperture area	1 m^2	-
View factor	0.8	-
Maximal heliostat aperture area	4751	m^2
Maximal net electric power	100	kW

In order to ensure that the ANN is properly trained, it is crucial that representative input/ output data are selected for training. The ANN architecture used in this study is illustrated in Figure 4. The optimal ANN architecture has been selected and trained. The data have been normalized and divided into two sets, 80 % for training and 20 % for test. For the selection of the best configuration, we must track the different steps represented in the flowchart of Figure 2 using a mathematical formula developed previously.

A variable number of neurons (from 3 to 10) were used in the hidden layer to observe any performance improvements obtained with the proposed modeling system. The graph of Figure 5 shows the variation of mean square error (MSE), defined in Equation (10), with the number of iteration during training of the network.

Table 2. Main exploitation characteristics of STPP

Parameters	Value	Unit
Inlet temperature of molten salt	290	^oC
Outlet temperature of molten salt	565	^oC
Inlet temperature of water	239	^oC
Outlet temperature of steam	552	^oC
Ambient temperature	20	^oC
Steam mass flow	0.3	kg / s
Beam radiation (DNI)	800	Wm^{-2}

Table 3. Statistical values of main simulated architectures

Architecture	SSE	RMSE %	R^2	MRE %
03-03-01	0.0119	1.2205	0.9976	3.1872
03-04-01	0.0088	1.0479	0.9981	2.2910
03-05-01	0.0119	1.2200	0.9974	2.9675
03-06-01	0.0122	1.2353	0.9975	3.1557

Table 3. (Continued)

Architecture	SSE	RMSE %	R^2	MRE %
03-07-01	0.0123	1.2399	0.9975	3.2800
03-08-01	0.0133	1.2889	0.9974	3.3679
03-09-01	0.0089	1.0543	0.9981	2.4800
03-10-01	0.0129	1.2711	0.9972	3.1331

We remark that the graph of simulation converges, after 100 iterations, to an error of about 10^{-6}.

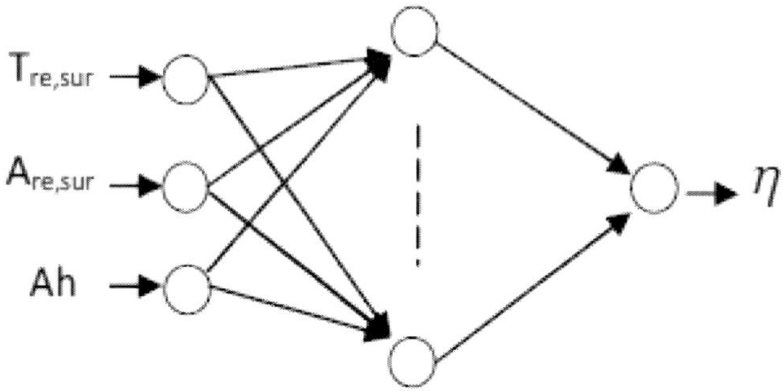

Figure 4. The architecture of ANN used in this study.

The results are reported in Figure 6 for the proposed ANN model of the receiver system. Statistical values such as SSE, RMSE, R^2 and MRE are given in Table 3, when the number of neurons in the hidden layer is varied from 3 to 10.

From the statistical data presented in Table 3, the architecture with four neurons in the hidden layer (03-04-01) appears to be the most optimal topology, because maximum R^2 value (99.81%) and minimum RMSE (1.0479), MRE (2.2910 %) and SSE (0.88 %) values were obtained.

As can be seen in Figure 6, the curves representing the calculated network output (observed efficiency) and the predicated network output (predicated efficiency) are close. This coincidence demonstrates the best training of the network.

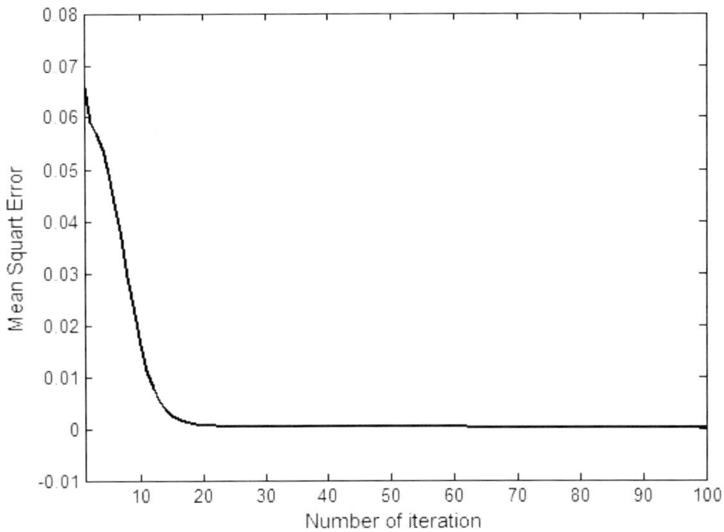

Figure 5. Training of the network (MSE).

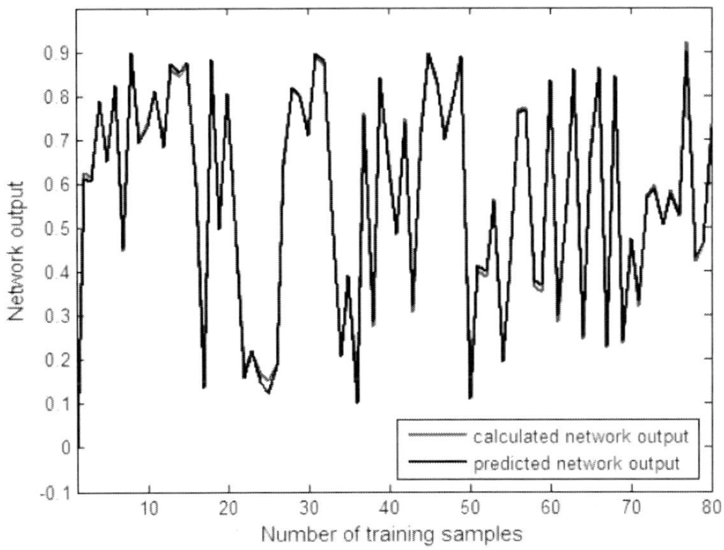

Figure 6. Comparison between actual and predicted efficiency.

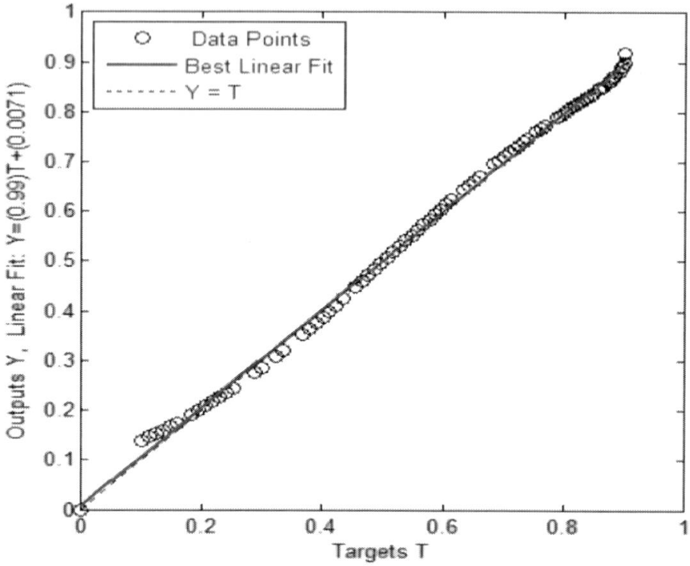

Figure 7. Performances of the optimal architecture.

In Figure 7, the dashed line is the perfect fit line where outputs and targets are equal to each other. The circles are the data points and continued line represents the best fit between outputs and targets. Here it is important to note that circles representing the network data are close to the dashed line. According to these results it can be deduced that the used algorithm of ANN predicts very well the thermal and the thermodynamic performance of solar tower receiver system.

It is not possible to know in advance the ideal network for an application. So every time a neural network (NN) is to be built in an application, it requires tests and experiments with different network settings or topologies to find a solution that performs well on the given application.

This is a problem because most NN requires a long training period – many iterations of the same pattern set. And even after much iteration, there is no way other than testing to see whether the network is efficiently mapping the training sets.

Figure 8 shows the testing of the optimal configuration (03-04-01). The network has been tested for 20 samples (20%).

A very good coincidence between the calculated network output (observed efficiency) and the predicated network output (predicated efficiency) could be noticed. This coincidence shows the best learning of the network.

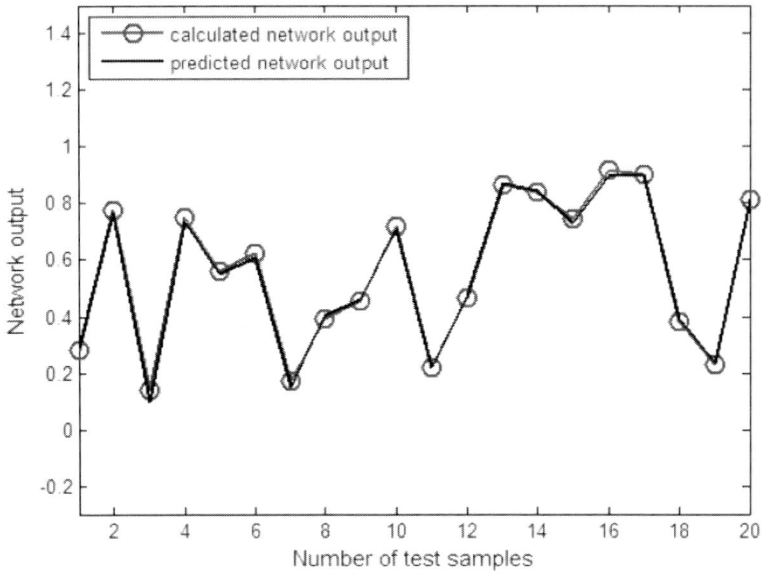

Figure 8. Testing of the network.

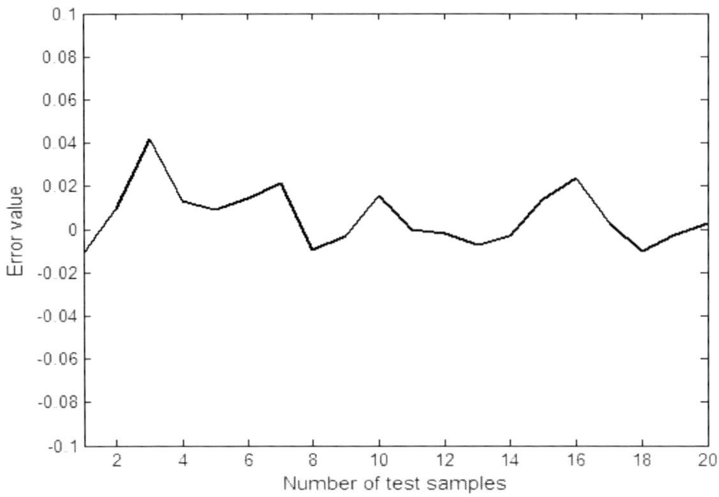

Figure 9. Error estimation between calculated and predicted test values.

Figure 9 represents the error estimation between the calculated and the predicted testing network for the optimal architecture. It is seen that the error values are between - 0.02 and + 0.04.

3.2.2. Overview of Main Performance Analysis Softwares

SOLARGY

The SOLERGY computer code was used to predict the annual energy production of the power plant designs. This code was selected because it has proven that it gives a realistic estimate of the actual energy produced by the Solar One and Solar Two plants.

SOLERGY simulates the operation of a solar central receiver power plant using solar irradiance data measured at 15-minute intervals. The relatively short intervals are needed to model plant startup and the effects of cloud transients.

The code has subroutines for each major plant system, i.e., heliostat field, receiver, thermal-energy storage, and turbine/generator. A program flow diagram is shown in Figure 10. For each 15-minute time step, SOLERGY determines the plant's operational state (shutdown, starting up, etc.) and calculates steady-state power flows through each plant system. Annual plant performance is found by summing the performance at every 15-minute time step.

SOLERGY's computational algorithms are based on simple conservation of energy.

There are no detailed thermodynamic calculations and no tracking of pressures and temperatures throughout the plant. Detailed calculations are necessary for the design and analysis of the performance over short periods but are not practical for an analysis of annual energy performance. Such calculations, however, are an important source of the data used as input to SOLERGY.

In SOLERGY, user-specified plant operational parameters are used to determine the plant's operation and performance during each time step. Examples of these parameters are (1) the time or energy required to start up and operate a system, (2) the parasitic load of major components and/or systems, and (3) the performance of individual components (e.g., receiver thermal losses or turbine efficiency).

Parameters are typically estimated from hand-done calculations or obtained by running complex computer codes outside of SOLERGY.

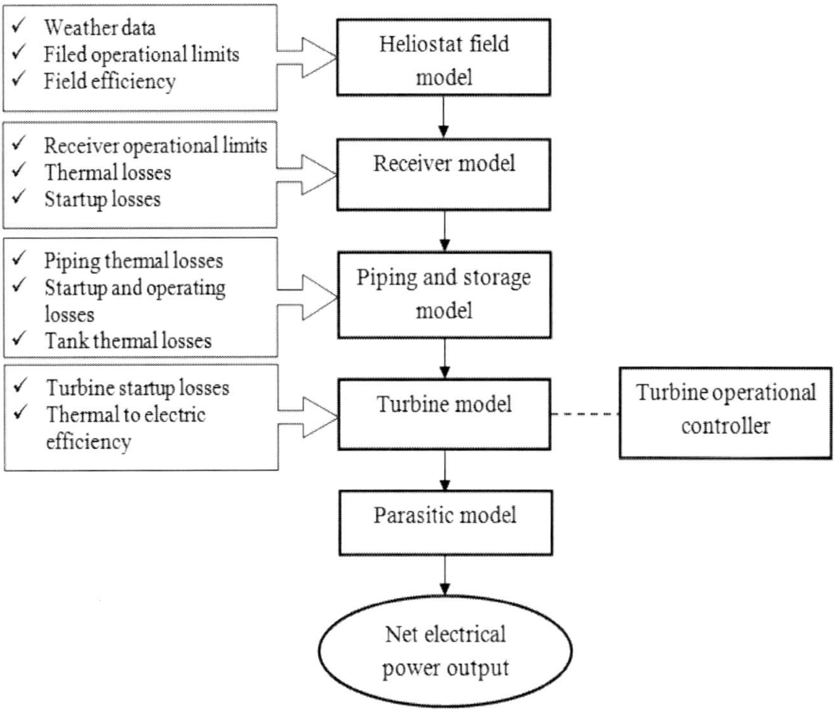

Figure 10. Program flow diagram for SOLERGY computer code.

TRNSYS

TRYNSYS is a software platform that enables the user to model different transient systems using modular components. Each component represents a physical process or feature in the system, and components can be developed and added, as needed, to a system model.

A component reads in a text-based input file and provides output through the solution of algebraic or differential equations. Components include solar thermal collectors (parabolic concentrating solar collector, flat plate solar collector), heat exchangers (counter-flow, cross-flow, parallel flow, shell-and-tube, waste heat recover, etc.), thermal storage tanks (stratified, variable volume, etc.), hydraulics (pumps, pipes, valves, etc.), controllers, and more.

Specific processes can be modeled for subcomponents of the total system, and total-system performance analyses can also be performed. The software contains a GUI that allows drag-and-drop arrangement and editing of component icons. Post-processing via graphing and reporting are also included in the software (Clifford, 2008).

DELSOL

DELSOL is a computer program developed by Sandia National Laboratories (US) for calculating collector field performance and layout and optimal system design for solar thermal central receiver plants. The code consists of a detailed model of the optical perfor-mance, a simpler model of the non-optical performance, an algorithm for field layout, and a searching algorithm to find the best system design based on energy cost.

The code can handle flat, focused and/or canted heliostats, and external cylindrical, multi-aperture cavity, and flat plate receivers. The program optimizes the tower height, receiver size, field layout, heliostat spacings, and tower position at user specified power levels subject to flux limits on the receiver and land constraints for field layout (Kistler, 1986).

SAM

SAM, originally called the "Solar Advisor Model," was first developed by the National Renewable Energy Laboratory (NREL) in collaboration with Sandia National Laboratories (SNL) in 2005 for internal use by the US Department of Energy's Solar Energy Technologies Program in its systems-based analysis of solar technology improvement opportunities within the program.

SAM is a computer model that evaluates performance and financial metrics of renewable energy systems (RES). SAM simulates the performance of photovoltaic, concentrating solar power, solar water heating, wind, geothermal, biomass, and conventional power systems. SAM's advanced simulation options facilitate parametric and sensitivity analyses, and statistical analysis capabilities are available for Monte Carlo simulation and weather variability studies. SAM can also read input variables from Microsoft Excel worksheets.

SAM's performance models make hour-by-hour calculations of a power system's electric output, generating a set of 8,760 hourly values that represent the system's electricity produc-tion over a single year.

It can explore the system's performance characteristics in detail by viewing tables and graphs of the hourly and monthly performance data, or use performance metrics such as the system's total annual output and capacity factor for more general performance evaluations (Blair et al., 2014).

4. Reliability Analysis

Reliability is the probability that unsatisfactory performance or failure will not occur. This definition is more adapted for the first phase of system life cycle (conception). The reliability can be defined during the system exploitation period as the probability that a system will perform its function over a specified period of time and under specified service conditions. To analyze the reliability of complex systems the analyst can follow two main techniques, the quantitative (Probabilistic Reliability Methods) and the qualitative (deter-ministic Reliability Methods) analysis.

The most commonly used quantitative techniques are Markov chain, Monte Carlo simulation, block diagram analysis, Time-Dependent (Hazard Functions), Response Surface Modeling and Reliability Index (β) Methods. While, the most used qualitative techniques are FMEA (Failure Modes and Effects Analysis), FMECA (Failure Modes, Effects and Criticality Analysis), EOE (Expert Opinion Elicitation), FTA (Fault Tree Analysis) and the traditional deterministic Stress-strength method.

In engineering design, the traditional deterministic design reliability model has been successfully applied to systematically reduce the failure probability and improve quality. However, the existence of uncertainties in either engineering simulations or manufacturing processes calls for a probabilistic reliability model (PRM) for reliable and safe designs.

The study of structural and mechanical reliability is concerned with the calculation and prediction of the probability of limit-state violations at any stage during a system's life. The probability of the occurrence of an event such as a limit-state violation is a numerical measure of the chance of its occurring. Once the probability is determined, the next goal is to choose design alternatives that improve system reliability and minimize the risk of failure.

In this part of the chapter, we present an overview of main reliability methods and tools which have been used to evaluate the reliability of RES, and then we focus our study only on the first order reliability method (FORM). Therefore, detailed analysis of FORM has been presented in this study. An example of hypothetical STPP has been provided to illustrate this method.

4.1. Main Probabilistic Characteristics of Reliability

Before treating a specific reliability problem, probabilistic key elements for reliability measurement are first presented succinctly. More details on the subject could be found in the literature (Villemeur, 1988; Ayyub and Mccuen, 1997; Birolini, 1997; Hoang, 2003).

4.1.1. Reliability Function or Survival Function

The reliability of a device after a time t corresponds to the probability that device has no failure between 0 and time t.

Denoting by T the random variable characterizing the moment of failure of the device, the reliability is expressed by the function R (t) such as: R (t) = Prob (that an entity E is not failing over time [0, t], assuming that is not faulty at instant t = 0)

$$R(t)=P(T\geq t)=1-F(t) \tag{15}$$

F (t) is the cumulative distribution function of the variable T.

Note that, the variable "time" is to be considered as a unit of use. Indeed, in the case of specific devices should be considered: a distance (km), number of laps, number of requests, etc.

Characteristic opposite of reliability (F (t)) is called probability of failure (P_f) or probability of unsatisfactory performance (P_{up}). It is the complement to 1 of reliability.

The probability of failure (P_f) is calculated during the exploitation phase of the system. It is easily defined for recurring events and replicate components (e.g., electronic and mechanical parts).

In the other hand the probability of unsatisfactory performance (P_{up}) is calculated during the conception phase of the system. It is nearly impossible to define for non-recurring events or unique components (e.g., sliding of gravity structures).

In the particular case when the study focuses on equipment operating in the solicitation (starter, air-bag, switch, ammunition), the measure of reliability is considered to be the probability that the equipment operates at the time of solicitation.

Figure 11. is a schematic representing the reliability function R (t), the cumulative distribution function F (t) and the probability density function *f* (t).

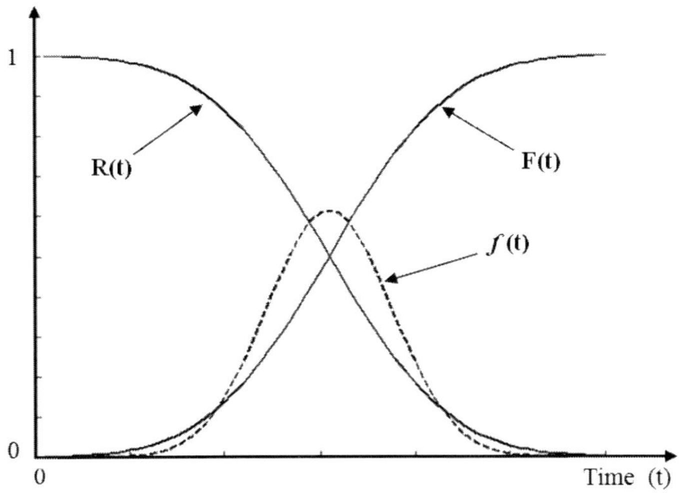

Figure 11. Reliability function R (t), cumulative distribution function F (t) and probability density function f (t).

4.1.2. Instantaneous Failure Rate

The failure rate, denoted by λ (t), of a device at time t is expressed mathematically by:

$$\lambda(t) = \lim_{\Delta t \to 0}\left(\frac{1}{\Delta t}\cdot\frac{R(t)-R(t+\Delta t)}{R(t)}\right) \tag{16}$$

Physically, the term λ (t) Δ t measures the probability that failure of a device occurs in the time interval [t, t + Δ t] knowing that this device worked well until time t.

The failure rate of a device at the time t is defined by

$$\begin{aligned}\lambda(t) &= -\frac{dR(t)}{dt}\cdot\frac{1}{R(t)}\\ &= \frac{dF(t)}{dt}\cdot\frac{1}{R(t)}\\ &= \frac{f(t)}{R(t)}\end{aligned} \tag{17}$$

$f(t)$ is the probability density function (pdf) of the time to failure.

The knowledge of the failure rate is sufficient to determine the reliability, using the following formula:

$$R(t) = \exp\left[-\int_0^t \lambda(s) \cdot ds\right] \tag{18}$$

4.1.3. Mean Time to Failure

The mean time to failure (MTTF) is the life expectancy T, is given by:

$$MTTF = E[T] = \int_0^{+\infty} t.f(t).dt = \int_0^{+\infty} R(u).du \tag{19}$$

By definition, the MTTF is the average lifespan of the system.

4.1.4. Mean Time to Repair

The Mean Time To Repair (MTTR) is a basic measure of the maintainability of repairable items. It represents the average time required to repair a failed component or device.

$$MTTR = \int_0^{\infty} t\ m(t)\ dt \tag{20}$$

where $m(t)$ is the repair time density probability function, it is given by:

$$m(t) = \mu(t)\ \exp\left[-\int_0^t \mu(\tau)\ d\tau\right] \tag{21}$$

where $\mu(t)$ is the repair rate.

4.1.5. Mean up Time

Mean up time (MUP) is the average operating time between the last restart after repair and the next failure. This time will be equal to the MTTF, if after each repair the system is "as good as new" (assimilation to a non-repairable system).

4.1.6. Mean Down Time

In organizational management, the mean down time (MDT) is the average time that a system is non-operational. This includes all downtime associated with repair, corrective and preventive maintenance, self-imposed downtime, and any logistics or administrative delays. The inclusion of delay times distinguishes the mean down time from the mean time to repair (MTTR), which includes only downtime specifically attributable to repairs (Smith, 2011).

4.1.7. Mean Time Between Failure

Mean time between failures (MTBF) is the predicted elapsed time between inherent failures of a system during operation (Jones, 2006). MTBF can be calculated as the arithmetic mean (average) time between failures of a system. The MTBF is typically part of a model that assumes the failed system is immediately repaired (mean time to repair, or MTTR), as a part of a renewal process. This is in contrast to the mean time to failure (MTTF), which measures average time to failures with the modeling assumption that the failed system is not repaired (infinite repair time). The MTBF can be calculated by the sum of MDT and MUP.

$$MTBF = MDT + MUP \tag{22}$$

Other reliability features that we have not mentioned are well described in any literature dealing with the subject of reliability: (Villemeur, 1988; Ayyub and Mccuen, 1997; Birolini, 1997; Hoang, 2003; Smith, 2011), etc.

Figure 12 shows the main average periods during the lifetime of material.

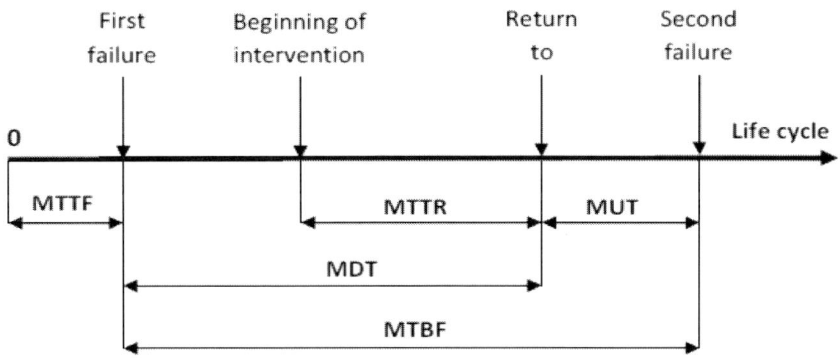

Figure 12. Average durations associated with reliability.

4.1.8. Availability

The notion of availability is related to repairable (or maintained) items only and is defined as the probability that a repairable system (or component) will function at time t. Respectively, the unavailability of a repairable system (or component) is defined as the probability that the item is in a failed state (down) at time t, which depends on the number of failures and repair times.

The availability A (t) is then given as the probability P that component E is not in a fai-lure state at time t:

A (t) = P [E not failed at t]

The availability may be expressed in terms of mean time between failures and the mean time to repair:

$$A_i = \frac{MTBF}{MTBF + MDT} \tag{23}$$

4.1.9. Maintainability

The maintainability is the ability of an entity to be maintained or restored on a given time in a state where it can perform a required function, when maintenance is performed under given conditions, with prescribed procedures and resources, It is characterized by the probabi-lity M(t) that the entity E is in state, at time t, to perform its functions, knowing that the entity was down at time 0.

M (t) = P [E is reparable on [0, t]]

The maintainability can be also expressed as the repair time distribution function by:

$$M(t) = 1 - \exp\left[-\int_0^t \mu(\tau)\, d\tau\right] \tag{24}$$

4.1.10. System Availability

For a series system with "n" independent components, system's availability is equal to the product of individual availabilities:

$$A(t) = \prod_{i=1}^{n} A_i(t) \tag{25}$$

where A_i (t) is the availability for each component.

The parallel configuration requires all components to be unavailable is order for the system to be unavailable:

$$A(t) = 1 - \prod_{i=1}^{n} \left[1 - A_i(t)\right] \tag{26}$$

4.2. Main Probability Distributions Used in Reliability Analysis

In this section, we present some distribution functions that occur most frequently in the material life cycle. We deal in particular with continuous distributions.

We will state the main properties of these laws (probability density, reliability and failure rate functions) and their application in reliability (Villemeur, 1988; Procaccia et al., 1992; Leemis, 1994; Ayyub and Mccuen, 1997; Birolini, 1997; Hoang, 2003).

4.2.1. Exponential Distribution

This law has many applications in several fields. It is a simple law, which is widely used in reliability when the failure rate is constant.

It describes the life of materials that undergo sudden failures.

The probability density function of an exponential distribution with parameter λ is written:

$$f(t) = \lambda e^{-\lambda t} \tag{27}$$

The reliability function:

$$R(t) = e^{-\lambda t} \tag{28}$$

The failure rate is constant over time:

$$\lambda(t) = \lambda \tag{29}$$

4.2.2. Normal Distribution (Laplace-Gauss)

The normal distribution is widespread among the laws of probability as it applies to many phenomena. In reliability, the normal distribution is used to represent the distribution of lifetimes of devices at the end of life (wear) because the failure rate is still increasing. It should only be used if the average duration of life is more than 3 times the standard deviation.

The probability density function (PDF) of a normal distribution with mean μ and standard deviation σ is written:

$$f(t) = \frac{1}{\sigma\sqrt{2\pi}} e^{-\frac{1}{2}\left(\frac{t-\mu}{\sigma}\right)^2} \tag{30}$$

The cumulative distribution function (CDF) is given as:

$$F(t) = \frac{1}{\sigma\sqrt{2\pi}} \int_{-\infty}^{t} e^{-\frac{(x-\mu)^2}{2\sigma^2}} dx \tag{31}$$

The reliability function is written by:

$$R(t) = 1 - \phi\left((t-\mu)/\sigma\right) \tag{32}$$

where ϕ is the cumulative distribution function (CDF) of the centered standard normal distribution (μ = 0) (σ = 1):

$$\phi(t) = \frac{1}{\sqrt{2\pi}} \int_{-\infty}^{t} e^{-\frac{u^2}{2}} du \tag{33}$$

4.2.3. Log-Normal Distribution (or Galton Law)

A positive continuous random variable T is distributed according to a log-normal distribution if its natural logarithm is distributed according to a normal distribution. This distribution is widely used to model life data, especially in mechanical fatigue failures.

The probability density function of a log-normal distribution with positive parameters μ and σ is:

$$f(t) = \frac{1}{\sigma \cdot t \cdot \sqrt{2\pi}} e^{-\frac{1}{2}\left(\frac{\log(t)-\mu}{\sigma}\right)^2} \quad , t > 0 \tag{34}$$

The reliability function is:

$$R(t) = 1 - \phi\left(\frac{\log(t) - \mu}{\sigma}\right) \tag{35}$$

$$R(t) = 1 - \frac{1}{\sigma\sqrt{2\pi}} \int_{-\infty}^{t} \frac{1}{x} e^{-\frac{1}{2}\left(\frac{\log(x)-\mu}{\sigma}\right)} dx \tag{36}$$

φ: Cumulative distribution function of the standard normal distribution.

The domain of definition is never negative. There is no limitation for the use of the log-normal distribution in reliability. The failure rate is increasing in the beginning of life and decreasing toward zero and the distribution is very asymmetrical.

4.2.4. Weibull Distribution

This is the most popular law, used in several fields (electronic, mechanical, etc.) (Chellali et al., 2010; Chellali et al., 2012). It allows to model in particular many wear material situations. It characterizes the behavior of the system in all three phases of life: youth period, period of useful life and period of wear or aging.

In its most general form, the Weibull distribution depends on the following three parameters: β, γ and η. The probability density function of a Weibull distribution is expressed by:

$$f(t) = \frac{\beta}{\eta}\left(\frac{t-\gamma}{\eta}\right)^{\beta-1} e^{-\left(\frac{t-\gamma}{\eta}\right)} \quad , t \geq \gamma \tag{37}$$

where: β is the shape parameter (β > 0), η is the scale parameter (η > 0) and γ is the position parameter (γ ≥ 0).

The reliability function is written as:

$$R(t) = e^{-\left(\frac{t-\gamma}{\eta}\right)^{\beta}} \tag{38}$$

The rate failure is given by:

$$\lambda(t) = \frac{\beta}{\eta}\left(\frac{t-\gamma}{\eta}\right)^{\beta-1}$$

(39)

Following the values of β, the failure rate is either decreasing (β <1), is constant (β = 1) or increasing (β > 1). The Weibull distribution can represent the three periods of the material life described by the bathtub curve (Chellali et al., 2010; Chellali et al., 2012).

If γ > 0 corresponds to materials whose failure probability is zero until a certain age γ.

4.2.5. Gamma Distribution

Gamma distribution is the law of the time of occurrence of α th event in a Poisson pro-cess. $\{T_i\}_{i=1,\alpha}$ is the vector representing the inter-event times (the time between successive failures of a system). If these times are independent and identically distributed random variables according to an exponential law of parameter β, the appearance cumulative time of α failures follows a Gamma distribution with parameter (α, β). Its probability density function is written:

$$f(t) = \frac{\beta^{\alpha} \cdot t^{\alpha-1} \cdot e^{-\beta t}}{\Gamma(\alpha)}, \quad t \geq 0,\ \alpha \geq 1 \text{ and } \beta \geq 0 \tag{40}$$

The failure rate is given by:

$$\lambda(t) = \frac{\beta^{\alpha} \cdot t^{\alpha-1} \cdot e^{-\beta t}}{\int_t^{\infty} \Gamma(\alpha) f(u) du} \quad (41)$$

4.2.6. Uniform Distribution

It is a law used in the Bayesian approach to model the expert advice for a given situation. The probability density function and cumulative function of a uniform law on [0, a] is given by the following formulas:

$$f(t) = \begin{cases} \frac{1}{a} & \text{if } 0 \le t \le a \\ 0 & \text{if } t > a \end{cases} \quad (42)$$

Generally, The probability density function of uniform distribution on [a, b] is written as:

$$f(t) = \begin{cases} \frac{1}{b-a} & \text{if } a \le t \le b \\ 0 & \text{otherwise} \end{cases} \quad (43)$$

The cumulative density function is given by:

$$F(t) = \begin{cases} 0 & \text{if } t \le a \\ \frac{t-a}{b-a} & \text{if } a \le t \le b \\ 1 & \text{if } t \ge a \end{cases} \quad (44)$$

4.3. Overview of Main Reliability Analysis Methods and Tools

As discussed previously, tow main techniques are used to analyze the reliability of complex systems; the probabilistic reliability methods (PRM) and deterministic reliability methods or non-probabilistic reliability methods (NPRM). The most probabilistic reliability methods used for renewable energy systems (RES) are Monte Carlo simulation (MCS), Markov chain and reliability indices. In the other hand, fault tree analysis (FTA), Failure modes effects and criticality analysis (FMECA) and reliability block diagram (RBD) are used as deterministic reliability methods to analyze RES.

The existing literature mostly focuses on reliability assessment for the photovoltaic (PV) systems, whereas much fewer references discuss the reliability evaluation for concentrating solar power CSP systems.

4.3.1. Monte Carlo Simulation

Monte Carlo simulation (MCS) has been largely used for reliability evaluation of RES. It treats problem as a series of experiment and simulates actual process and random behaviour of system under consideration (Paliwal et al., 2014). Conceptually, Monte-Carlo simulation owns much more flexibility and can be used to study more complicated problems, such as reliability assessment of PV-Wind hybrid (Saheb-Koussa et al., 2009; Kishore and Fernandez, 2011) system and CSP systems. Ravindra and Kulkarni (Moharil and Kulkarni, 2010) have developed MATLAB program based on Monte Carlo simulation technique (MCS) for reliability analysis of small isolated power system using solar photovoltaic (SPV). Various reliability indices are obtained by HMSR method using MCS technique. Reliability indices, additional load hours (ALH) and additional power (AP) reduce exponentially with increase in load indicates that a SPV source will offset maximum fuel when all of its generated energy is utilized.

Ho et al. (Ho, 2010) have developed a probabilistic performance model of concentrating solar power plants including quantification of uncertainties inherent in the system. Monte Carlo or Latin hypercube sampling (LHS) methods are then implemented in the model to generate many different (but equally probable) realizations of the system performance.

4.3.2. Fault Tree Analysis

The fault tree technique was introduced in 1962 at Bell Telephone. Laboratories, in connection with a safety evaluation of the launching system for the intercontinental Minute-man missile.

The Boeing Company improved the technique and introduced computer programs for both qualitative and quantitative fault tree analysis. Today fault tree analysis is one of the most commonly used techniques for risk and reliability studies.

A FTA can yield both qualitative and quantitative information about the system under study. Qualitative information may include failure paths, root causes, and weak areas of the system. Quantitative analysis of a fault tree gives a probabilistic estimation of the top event and can lead to a conclusion as to whether the design is adequate in terms of reliability and safety.

In the field of RES, many scholars have used the FTA for the evaluation of reliability systems (Wu et al., 2011; Zini et al., 2011). Wu et al. (Wu et al., 2011) have used the fault tree analysis (FTA) and fuzzy reasoning Petri net (FRPN) models to analyze the reliability of a solar array mechanical system in order to find mechanisms of the solar array fault. The analysis results show that harsh thermal environment and impact caused by particles in space are the most vital causes of the solar array fault. Zini et al. (Zini et al., 2011) have proposed a method based on FTA for assessing the reliability of large-scale grid-connected photovoltaic systems.

4.3.3. Markov Chain

There are two basic Markov analysis methods: Markov Chain and Markov Process. The Markov Chain assumes discrete states and a discrete time parameter; With the Markov Process, states are continuous. A Markov chain may be described as Homogeneous or Non-homogeneous.

A Homogeneous Markov Chain is characterized by constant transition rates between the states. A Non-homogeneous Markov Chain is characterized by the fact that the transition rates between the states are functions of a global clock e.g., elapsed mission time.

Many investigations have been carried out for PV system reliability using a Markov chain method (Urbina et al., 1998; Garro and Barrara, 2011). Ehnberg (Ehnberg, 2003) has proposed two stochastic models, based on Markov theory, in order to evaluate the reliability of isolated power systems with solar, wind and hydro generation.

4.3.4. Reliability Index Methods

Uncertainties are involved in every part of a solar thermal system due to tolerances in design parameters, tolerances in system parts, and the random nature of variables like solar radiation and thermal demand.

To estimate the level of uncertainty in the behavior of solar mechanical and thermal systems, the reliability index methods (RIM) are used. The most RIM used to assess the design reliability and safety are: the first order reliability method (FORM) and the second order reliability method (SORM).

Many references, concerning the application of FORM and SORM for the structural and mechanical reliability, have been published, whereas no much work has discussed the use of these methods for the reliability assessment of

RES. Do et al. (Do et al., 2011) have proposed to use the First-Order Reliability Method (FORM), coupled with the law of total probability, for short-term reliability assessment of power systems with large amount of wind power. They (Do et al., 2011) have concluded that this method is much faster than the MCS. More details about FORM are reported subsequently.

4.3.5. Reliability Block Diagram

A reliability block diagram (RBD) is a success-oriented network describing the function of the system. It shows the logical connections of (functioning) components needed to fulfill a specified system function. In a reliability block diagram, components are symbolized by rectangular blocks, which are connected by straight lines according to their logic relationship. Depending on the purpose of the system analysis, a block may represent a lowest-level component, a module, or a subsystem. Collins et al. (Collins et al., 2009) have used RBD to develop the photovoltaic reliability and performance model (PV-RPM). Garro and Barrara (Garro and Barrara, 2011) have proposed reliability model, for a residential photovoltaic system, based on SERA (Software Environment for Reliability Analysis).

The reliability results obtained from SERA have been compared with those produced by applying the RDB model.

4.3.6. Failure Modes, Effects and Criticality Analysis

Failure modes effects and criticality analysis (FMECA) aims to assess the effects of possible failures and correlate them to the failure rate of occurrence. This risk assessment provides two steps: (i) preparation of FMECA worksheets, in which critical events (failure modes) are identified and the effects of them on the global system are evaluated following a standard worksheet format; (ii) risk priority numbers evaluation, in which the risk associated with each failure mode is assessed by using the formula RPN = S*O*D, where S (Severity) is a numerical subjective estimate of how severe is the effect of a failure mode, O (Occurrence) is the rate of occurrence of the failure mode, D (Detection) is a numerical subjective estimate of the effectiveness of the controls to prevent or detect the cause or failure mode before the failure happens. The evaluation of RPN leads to the definition of a risk matrix which gives a compact and visual representation of the severity of the risk of the considered critical events. A review on the past studies shows that FMECA technique is used in some fields of power systems, e.g. wind turbines and solar modules (Akbari et al., 2013).

4.3.7. Pro-Opta Reliability Analysis Software

SNL's Pro-Opta software is a reliability analysis and optimization suite that provides an analyst with the information needed to make strategic decisions in the competitive market-place. The tool offers the flexibility to create a system's reliability model using either field data (maintenance records) or summarized data, e.g., from a FMECA. Once the model is created, decisions can be made on which components should be improved, based on sensiti-vity analysis. Potential spares and reliability improvement options can be assessed and ranked to aid in evaluation of alternatives based on availability, reliability, and cost impacts.

The Pro-Opta software was used to predict plant unavailability. This software has many capabilities that allow a detailed evaluation of the reliability of existing power plants as well as the optimization of proposed future plant designs. The package is currently being used to evaluate and improve the reliability of wind-turbine power plants.

To run the software, the analyst must provide a logic model and reliability data for the power plant components and/or subsystems. The logic model (called a "fault tree" or "reliability block diagram") relates the series-parallel arrangement of the plant components. Two pieces of reliability data are input for each plant component: (1) the mean time between failures (MTBF), and (2) the mean downtime (MDT). Given this information, component availability is calculated from the following equation:

$$A = MTBF/(MTBF + MDT).$$

The numerator of this equation represents the hours the component is available for service, and the denominator represents total clock hours. The component availabilities for the individual components are combined via the logic model to calculate the availability for the total power plant. In addition to plant availability, the code provides other figures of merit regarding the plant.

Pro-Opta Enables the:

- Analysis of field maintenance data and computation of system metrics;
- Identification of key drivers of system unreliability;
- Prediction of equipment lifecycle performance;
- Assessment of modifications to maintenance practices;
- Ranking of potential reliability improvement and inventory options;
- Modeling system reliability using graphical fault tree;

- Evaluation of fleets with multiple configurations;
- Trade-offs of cost versus performance.

As an example of applications, Pro-Opta can be used in the:

- Evaluation of energy efficiency improvements;
- Device reliability predication;
- Modernization upgrades impact analysis;
- Sustainment assessment;
- Fleet reliability planning;
- Defense, energy, aviation, and commercial;
- Industries.

4.4. First Order Reliability Method (FORM)

FORM, abbreviation for the first-order reliability method, remedies the invariance problem of mean value first-order second moment (MVFOSM) described subsequently. The remedy is to approximate the limit-state function somewhere on the limit-state surface, i.e., at a point where g=0 instead of at the mean. The limit-state surface, which separates the failure domain from the safe domain, is shared by all equivalent limit-state functions.

4.4.1. First-Order Second Moment (FOSM) Method

The FOSM method, also referred to as the Mean Value FOSM (MVFOSM), simplifies the functional relationship and alleviates the complexities of the probability-of-failure calculation. The name “first-order” come from the first-order expansion of the function. As implied, inputs and outputs are expressed as the mean and standard deviation. Higher moments, which might describe skew and flatness of the distribution, are ignored.

In the MVFOSM method, the limit-state function is represented as the first order Taylor series expansion at the mean value point. Assuming that the variables X are statistically independent, the approximate limit-state function at the mean is written as:

$$g_a(X) \approx g(\mu_X) + \nabla g(\mu_X)^T (X_i - \mu_{X_i}) \tag{45}$$

where, $\mu_X = \left\{\mu_{x_1}, \mu_{x_2}, \ldots \mu_{x_n}\right\}^T$ is the mean of the basic random variable *X*, and $\nabla g(\mu_X)$ is the gradient of *g* evaluated at μ_x,

$$\nabla g(\mu_X) = \left\{\frac{\partial g(\mu_X)}{\partial x_1}, \frac{\partial g(\mu_X)}{\partial x_2}, \ldots, \frac{\partial g(\mu_X)}{\partial x_n}\right\}^T$$

The mean value of approximate limit-state function *g(X)* is:

$$\mu_{g_a} \approx E\left[g(\mu_X)\right] = g(\mu_X) \tag{46}$$

Therefore, the standard deviation of the approximate limit-state function is:

$$\sigma_{g_a} = \sqrt{Var\left[g_a(X)\right]} = \sqrt{\left[\nabla g(\mu_X)^T\right]^2 Var(X)}$$
$$= \left[\sum_{i=1}^{n}\left(\frac{\partial g(\mu_X)}{\partial x_i}\right)^2 \sigma_{x_i}^2\right]^{\frac{1}{2}} \tag{47}$$

where, σ_{x_i} is the standard deviation of the basic random variable *X*.

The reliability index β is computed as:

$$\beta = \frac{\mu_{g_a}}{\sigma_{g_a}} \tag{48}$$

whereas the probability of failure can be obtained as:

$$P_f = \Phi(-\beta) \tag{49}$$

where Φ is the cumulative distribution function of the standard normal variable.

4.4.2. Hasofer and Lind (HL) Safety-Index

Searching for the MPP on the limit-state surface is a key step in the HL method. The improvement of the HL method compared with the MVFOSM also comes from changing the expansion point from the mean value point to the MPP.

The safety-index β is the shortest distance from the origin to the failure surface $g(U) = 0$, i.e.,

$$\beta = \min_{U \in g(U)=0} \left(U^T U\right)^{\frac{1}{2}} \tag{50}$$

This safety-index is also called the Hasofer and Lind (HL) safety-index, β_{HL}. The point $U^*\left(u_1^*, u_2^*, ..., u_n^*\right)$ on $g\ (U) = 0$ is the design point, the values-of-safety, where, U is the standard normal variable with the mean equal to zero and the standard deviation equal to one.

$$U = \frac{\left(X - \mu_X\right)}{\sigma_{x_i}} \tag{51}$$

Equation 50 shows that safety-index, β, is the solution of a constrained optimization problem in the standard normal space.

Minimize: $\beta(U) = \left(U^T U\right)^{\frac{1}{2}}$ (52)

Subject to: $g(U) = 0$ (53)

There are many algorithms available that can solve this problem, such as mathematical optimization schemes or other iteration algorithms. In the following description, the most commonly used recursive algorithms, the HL and HL-RF methods, are introduced to solve the reliability problems. The HL method was proposed by Hasofer and Lind. Rackwitz and Fiessler extended the HL method to include random variable distribution information, calling their extended method the HL-RF method. Assuming that the limit-state surface with n-dimensional normally distributed and inde-pendent random variables X is:

$$g(X) = g\left(\{x_1, x_2, ..., x_n\}^T\right) = 0 \tag{54}$$

This limit-state function can be linear or nonlinear. The limit-state function given in Equation 54 is transformed into:

$$g(U) = g\left(\left\{\sigma_{x_1}u_1 + \mu_{x_1}, \sigma_{x_2}u_2 + \mu_{x_2}, ..., \sigma_{x_n}u_n + \mu_{x_n}\right\}^T\right) = 0 \tag{55}$$

The normal vector from the origin ô to the limit-state surface g (U) generates an intersection point P^* as shown in Figure 13. The distance from the origin to the MPP is the safety-index β. The first-order Taylor series of expansion of g (U) at the MPP U^* is:

$$g_a(U) \approx g(U^*) + \sum_{i=1}^{n} \frac{\partial g(U^*)}{\partial U_i}(u_i - u_i^*) \tag{56}$$

The shortest distance from the origin to the above approximate failure surface given in Equation 56 is:

$$\hat{O}P^* = \beta = \frac{g(U^*) - \sum_{i=1}^{n} \frac{\partial g(U^*)}{\partial x_i}\sigma_{x_i}u_i^*}{\sqrt{\sum_{i=1}^{n}\left(\frac{\partial g(U^*)}{\partial x_i}\sigma_{x_i}\right)^2}} \tag{57}$$

The direction cosine of the unit outward normal vector is given as:

$$\cos\theta_{x_i} = \cos\theta_{u_i} = -\frac{\dfrac{\partial g\left(U^*\right)}{\partial u_i}}{\left|\nabla g\left(U^*\right)\right|}$$

$$= -\frac{\dfrac{\partial g\left(X^*\right)}{\partial x_i}\sigma_{x_i}}{\left[\sum_{i=1}^{n}\left(\dfrac{\partial g\left(X^*\right)}{\partial x_i}\sigma_{x_i}\right)^2\right]^{\frac{1}{2}}} = \alpha_i \tag{58}$$

where α_i expresses the relative effect of the corresponding random variable on the total variation. Thus, it is called the *sensitivity factor*.

The coordinates of the point P^* are computed as:

$$u_i^* = \frac{x_i^* - \mu_{x_i}}{\sigma_{x_i}} = \hat{O}P^* \cos\theta_{x_i} = \beta\alpha_i \tag{59}$$

The coordinates corresponding to P^* in the original space are

$$x_i^* = \mu_{x_i} + \beta\sigma_{x_i}\alpha_i, \quad \left(i = 1,2,\ldots,n\right) \tag{60}$$

Since P^* is a point on the limit-state surface,

$$g\left(\left\{x_1^*, x_2^*, \ldots x_n^*\right\}^T\right) = 0 \tag{61}$$

4.4.3. Hasofer Lind - Rackwitz Fiessler (HL-RF) Method

The HL algorithm could consider normal random variables for the determination of the MPP point. However, many structural reliability problems involve non-normal random variables. There are many methods available for conducting the transformations, such as Rosenblatt (Rackwitz and Fiessler, 1978; Santosh et al., 2006). Rackwitz and Fiessler (1978) extended algorithm developed by the Hasofer and Lind algorithm in order to include the distribution information of the basic random variables. The extended algorithm could consider non-normal basic random variables with a simple and approximate transformation. Hence, Equation 51 can be written as follows (Rackwitz and Fiessler, 1978; Santosh et al., 2006; Keshtegar and Miri, 2013):

$$U = \frac{\left(X - \mu_{x_i}^e\right)}{\sigma_{x_i}^e} \tag{62}$$

The equivalent mean $\mu_{x_i}^e$ and the standard deviation $\sigma_{x_i}^e$ are given as (Rackwitz and Fiessler, 1978; Santosh et al., 2006; Choi et al., 2007):

$$\sigma_{x_i}^e = \frac{\phi\left(\Phi^{-1}\left[F_{x_i}\left(x_i^*\right)\right]\right)}{f_{x_i}\left(x_i^*\right)} \tag{63}$$

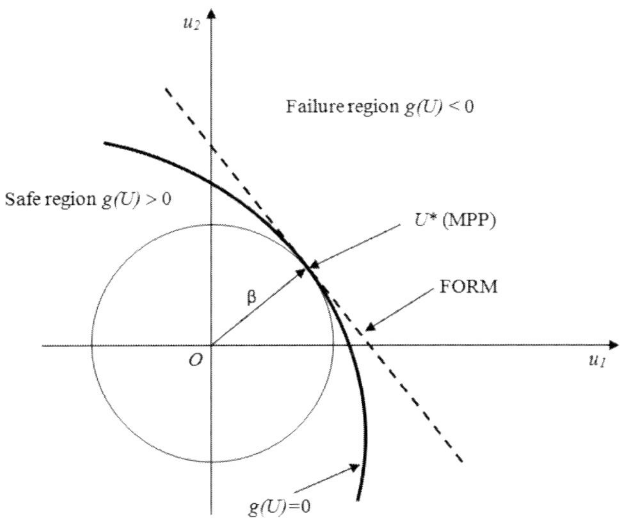

Figure 13. Illustration of the limit-state function and the safety index.

$$\mu_{x_i}^e = x_i^* - \Phi^{-1}\left[F_{x_i}\left(x_i^*\right)\right]\sigma_{x_i}^e \tag{64}$$

where, $f_{x_i}\left(x_i^*\right)$ and $F_{x_i}\left(x_i^*\right)$ are the probability density function and the marginal cumula-tive function respectively, at the MPP point.

The main steps at the HL-RF iteration method are:

1) Define the appropriate limit-state function of Equation 54
2) Compute of the mean and standard deviation of the equivalent normal variables based on Equations 52 and 53.
3) Set the mean value point as an initial design point, i.e, $x_{i,k} = \mu_{x_i}$ $i=1,2,...,n,$ and compute the gradient $\nabla g(X_k)$ of the limit-stat function at this point. Here $x_{i,k}$ *is* the ith element in the vector X_k of the kth iteration.
4) Compute the initial β using the mean-value method (Cornell safety-index), i.e., $\beta = \mu_{g_a} / \sigma_{g_a}$ and its direction cosine.
5) Compute of the mean and standard deviation of the equivalent normal variables based on Equations Equation 52 and 53.
6) Compute a new design point X_k and U_k (Equations 59 and 60), function value, and gradients at this new design point.
7) Compute the safety index β using Equation 57 and the direction cosine or sensitivity factor from Equation 58
9) Compute the coordinates of the design point X_k or most probable failure point (MMP), X^*.

4.5. Example: Reliability Analysis of STPP

High value of the receiver unsatisfactory performance factor (F_{up}) indicates the existence of failure in STPP design. Therefore, equipment malfunctions will cause a solar power tower to be unavailable for power production during a certain fraction of the calendar year. In this example the reliability of the receiver design is analyzed in order to estimate the probability of failure related to the unsatisfactory performance factor. Proposition of new design is identified and a solution to increase the reliability of receiver design is also discussed.

The example of the STPP taken into account for the reliability analysis is the same as that considered for the performance analysis. Table 1 and 2 show the main STPP system characteristics. In this study we use the first order reliability method (FORM) to analyze the reliability and probability of failure of STPP. This analysis has been based on the main HL-RF algorithm steps presented previously. The limit state function is obtained using the modified analytical model from (Benammar et al., 2014). The losses in original model have then been represented by the unsatisfactory performance factor (F_{up}). F_{up} is the receiver incident energy fraction loss. In the HL-RF algorithm cited below, F_{up} is equal to 0.1 or to 10%. We assume in this example that the distribution functions of all variables is uniform, so we do not need to compute the mean and standard deviation of the equivalent normal variables.

4.5.1. HL-RF Algorithm Procedure for STPP System

Iteration (1)

1) Define the appropriate limit-state function of Equation 54

$$g(X) = \eta_{STPP} \times \left(K_{Tre,sur} T_{re,sur} + K'_{Tre,sur} \right) - \dot{m}_{st} \left(Cp_{st,5} T_5 - Cp_{st,4} T_4 \right)$$

where

$$X = \left\{ T_{re,sur}, \dot{m}_{st1}, A_{re,sur}, \lambda_{tube} \right\}$$

$\eta_{STPP} = 1 - F_{up}$, is the solar tower power plant receiver efficiency

F_{up} is the unsatisfactory performance factor, (here, F_{up} varies between 0.1 and 0.9)

$$K_{Tre,sur} = \frac{A_{re,sur}}{\dfrac{d_o}{d_i h_{ms}} + \dfrac{d_o}{2\lambda_{tube}} \ln \dfrac{d_o}{d_i}}$$

$$K'_{Tre,sur} = \frac{T_{ms} A_{re,sur}}{\dfrac{d_o}{d_i h_{ms}} + \dfrac{d_o}{2\lambda_{tube}} \ln \dfrac{d_o}{d_i}}$$

$T_{re,sur}$ is the receiver surface temperature

$\dot{m}_{st}$ is the steam mass flow

$A_{re,sur}$ is the receiver surface area

λ_{tube} is the tube conductivity

d_i and d_o Are respectively the inlet and outlet tube diameter

T_{ms} *and* h_{ms} Are respectively the molten salt temperature and the convective coefficient

$C_{p\,4}$ *and* C_{p5} Are respectively the inlet and outlet specific heat capacity of steam in the steam generator.

2) Set the mean value point as an initial design point and compute the gradient $\nabla g\left(X_1\right)$ of the limit-stat function at $X_1 = \left\{T_{re,sur1}, \dot{m}_{st1}, A_{re,sur1}, \lambda_{tube1}\right\}$.

$$T_{re,sur1} = \mu_{T_{re,sur1}} = 600\ ^{\circ}C$$

$$\dot{m}_{st1} = \mu_{\dot{m}_{st1}} = 3 \text{ kg/s}$$

$$A_{re,sur1} = \mu_{A_{re,sur1}} = 1 \text{ m}^2$$

$$\lambda_{tube1} = \mu_{\lambda_{tube1}} = 23.9 \text{ W/m K}$$

$$\sigma_{T_{re,sur}} = 10\ \%$$

$$\sigma_{\dot{m}_{st}} = 10\ \%$$

$$\sigma_{A_{re,sur}} = 15\ \%$$

$$\sigma_{\lambda_{tube}1} = 5\ \%$$

The gradients $\nabla g\left(X_k\right)$ of the limit-stat function at the design point are computed as:

$$\nabla g\left(T_{re,sur1}\right)=\mathrm{F_{up}}\times\frac{A_{re,sur1}}{\dfrac{d_o}{d_i h_{ms}}+\dfrac{d_o}{2\lambda_{tube1}}\ln\dfrac{d_o}{d_i}}=1.1938\ 10^4$$

$$\nabla g\left(\dot{m}_{st1}\right)=-\ Cp_{st,5}T_5-Cp_{st,4}T_4=\text{-}2.3415\ 10^6$$

$$\nabla g\left(A_{re,sur1}\right)=\mathrm{F_{up}}\times\frac{T_{re,sur1}+T_{ms}}{\dfrac{d_o}{d_i h_{ms}}+\dfrac{d_o}{2\lambda_{tube1}}\ln\dfrac{d_o}{d_i}}=1.2236\ 10^7$$

$$\nabla g\left(\lambda_{tube1}\right)=\mathrm{F_{up}}\times\frac{A_{re,sur1}\left(T_{re,sur1}+T_{ms}\right)\left(\dfrac{d_0}{2}\ln\dfrac{d_0}{d_i}\right)}{\left(\dfrac{d_0}{d_i\,\mathrm{h}_{ms}}\lambda_{tube1}+\dfrac{d_0}{2}\ln\dfrac{d_0}{d_i}\right)^2}=5.0736\ 10^5$$

3) Compute the initial β using the mean-value method (Cornell safety-index), i.e., $\beta=\mu_{g_a}/\sigma_{g_a}$ and its direction cosine.

$$\beta_1=\frac{\mu_{g_a}}{\sigma_{g_a}}=\frac{g\left(X_1\right)}{\sqrt{\left(\dfrac{\partial g\left(X_1\right)}{\partial T_{re,sur}}\sigma_{T_{re,sur}}\right)^2+\left(\dfrac{\partial g\left(X_1\right)}{\partial\dot{m}_{st}}\sigma_{\dot{m}_{st}}\right)^2+\left(\dfrac{\partial g\left(X_1\right)}{\partial A_{re,sur}}\sigma_{A_{re,sur}}\right)^2+\left(\dfrac{\partial g\left(X_1\right)}{\partial\lambda_{tube}}\sigma_{\lambda_{tube}}\right)^2}}=2.3425$$

$$\alpha_{T_{re,sur1}}=-\frac{\left.\dfrac{\partial g\left(X_1\right)}{\partial T_{re,sur}}\right|_{\mu_{T_{re,sur}}}\sigma_{T_{re,sur}}}{\sqrt{\left(\dfrac{\partial g\left(X_1\right)}{\partial T_{re,sur}}\sigma_{T_{re,sur}}\right)^2+\left(\dfrac{\partial g\left(X_1\right)}{\partial\dot{m}_{st}}\sigma_{\dot{m}_{st}}\right)^2+\left(\dfrac{\partial g\left(X_1\right)}{\partial A_{re,sur}}\sigma_{A_{re,sur}}\right)^2+\left(\dfrac{\partial g\left(X_1\right)}{\partial\lambda_{tube}}\sigma_{\lambda_{tube}}\right)^2}}=\text{- }0.3289$$

$$\alpha_{\dot{m}_{st1}} = -\frac{\left.\frac{\partial g(X_1)}{\partial \dot{m}_{st}}\right|_{\mu_{\dot{m}_{st}}} \sigma_{\dot{m}_{st}}}{\sqrt{\left(\frac{\partial g(X_1)}{\partial T_{re,sur}}\sigma_{T_{re,sur}}\right)^2 + \left(\frac{\partial g(X_1)}{\partial \dot{m}_{st}}\sigma_{\dot{m}_{st}}\right)^2 + \left(\frac{\partial g(X_1)}{\partial A_{re,sur}}\sigma_{A_{re,sur}}\right)^2 + \left(\frac{\partial g(X_1)}{\partial \lambda_{tube}}\sigma_{\lambda_{tube}}\right)^2}} = 0.3225$$

$$\alpha_{A_{re,sur1}} = -\frac{\left.\frac{\partial g(X_1)}{\partial A_{re,sur}}\right|_{\mu_{A_{re,sur}}} \sigma_{A_{re,sur}}}{\sqrt{\left(\frac{\partial g(X_1)}{\partial T_{re,sur}}\sigma_{T_{re,sur}}\right)^2 + \left(\frac{\partial g(X_1)}{\partial \dot{m}_{st}}\sigma_{\dot{m}_{st}}\right)^2 + \left(\frac{\partial g(X_1)}{\partial A_{re,sur}}\sigma_{A_{re,sur}}\right)^2 + \left(\frac{\partial g(X_1)}{\partial \lambda_{tube}}\sigma_{\lambda_{tube}}\right)^2}} = -0.8428$$

$$\alpha_{\lambda_{tube1}} = -\frac{\left.\frac{\partial g(X_1)}{\partial \lambda_{tube}}\right|_{\mu_{\lambda_{tube}}} \sigma_{\lambda_{tube}}}{\sqrt{\left(\frac{\partial g(X_1)}{\partial T_{re,sur}}\sigma_{T_{re,sur}}\right)^2 + \left(\frac{\partial g(X_1)}{\partial \dot{m}_{st}}\sigma_{\dot{m}_{st}}\right)^2 + \left(\frac{\partial g(X_1)}{\partial A_{re,sur}}\sigma_{A_{re,sur}}\right)^2 + \left(\frac{\partial g(X_1)}{\partial \lambda_{tube}}\sigma_{\lambda_{tube}}\right)^2}} = -0.2784$$

3) Compute a new design point X_1 and U_1 (Equations 59 and 60), function value, and gradients at this new design point.

The coordinates of the new design point in the original space using Equation 60:

$$T_{re,sur2} = \mu_{T_{re,sur}} + \beta_1 \sigma_{T_{re,sur}} \alpha_{T_{re,sur1}} = 553.7743\ ^\circ\text{C}$$

$$\dot{m}_{st2} = \mu_{\dot{m}_{st}} + \beta_1 \sigma_{\dot{m}_{st}} \alpha_{\dot{m}_{st}} = 3.2267\ \text{kg/s}$$

$$A_{re,sur2} = \mu_{A_{re,sur}} + \beta_1 \sigma_{A_{re,sur}} \alpha_{A_{re,sur}} = 0.7039\ \text{m}^2$$

$$\lambda_{tube2} = \mu_{\lambda_{tube}} + \beta_1 \sigma_{\lambda_{tube}} \alpha_{\lambda_{tube}} = 23.1207\ \text{W/m K}$$

The coordinates of the new design point in the normal standard space using Equation 59:

$$u_{T_{re,sur}2} = \frac{T_{re,sur2} - \mu_{T_{re,sur}}}{\sigma_{T_{re,sur}}} = -0.7704$$

$$u_{\dot{m}_{st}2} = \frac{\dot{m}_{st2} - \mu_{\dot{m}_{st}}}{\sigma_{\dot{m}_{st}}} = 0.7556$$

$$u_{A_{re,sur}2} = \frac{A_{re,sur2} - \mu_{A_{re,sur}}}{\sigma_{A_{re,sur}}} = \text{-}1.9742$$

$$u_{\lambda_{tube}2} = \frac{\lambda_{tube2} - \mu_{\lambda_{tube}}}{\sigma_{\lambda_{tube}}} = \text{- }0.6522$$

The probability of failure is computed using Equation 49 as:

$$P_f = \Phi(-\beta_1) = 0.0096$$

Iteration (2)

1) Compute the function value and gradients of the limit-state function at

$$X_2 = \left\{T_{re,sur2}, \dot{m}_{st2}, A_{re,sur2}, \lambda_{tube2}\right\}$$

$$g(X_2) = \eta_{STPP} \times \left(K_{Tre,sur}\, T_{re,sur2} + K'_{Tre,sur}\right) - \dot{m}_{st2}\left(Cp_{st,5}T_5 - Cp_{st,4}T_4\right) = \text{- }0.7429{\times}10^7$$

The gradients $\nabla g(X_2)$ of the limit-stat function at X_2 are computed as:

$$\nabla g(T_{re,sur2}) = \mathrm{F_{up}} \times \frac{A_{re,sur2}}{\dfrac{d_o}{d_i h_{ms}} + \dfrac{d_o}{2\lambda_{tube}} \ln\dfrac{d_o}{d_i}} = 249.6432$$

$$\nabla g(\dot{m}_{st2}) = -\ Cp_{st,5}T_5 - Cp_{st,4}T_4 = \text{-}2.3782\ 10^6$$

$$\nabla g(A_{re,sur2}) = \mathrm{F_{up}} \times \frac{T_{re,sur2} + T_{ms}}{\dfrac{d_o}{d_i h_{ms}} + \dfrac{d_o}{2\lambda_{tube}} \ln\dfrac{d_o}{d_i}} = 3.4715\ 10^5$$

$$\nabla g\left(\lambda_{tube2}\right) = \mathrm{F_{up}} \times \frac{A_{re,sur2}\left(T_{re,sur2} + T_{ms}\right)\left(\frac{d_0}{2}\ln\frac{d_0}{d_i}\right)}{\left(\frac{d_0}{d_i\ \mathrm{h}_{ms}}\lambda_{tube2} + \frac{d_0}{2}\ln\frac{d_0}{d_i}\right)^2} = 3.4706\ 10^5$$

2) Compute the initial β using Equation 57 and the direction cosine α_i:

$\beta_2 =$

$$= \frac{g(X_2) - \frac{\partial g(X_2)}{\partial T_{re,sur}}\sigma_{T_{re,sur}} u_{T_{re,sur}2} - \frac{\partial g(X_2)}{\partial \dot{m}_{st}}\sigma_{\dot{m}_{st}} u_{\dot{m}_{st}2} - \frac{\partial g(X_2)}{\partial A_{re,sur}}\sigma_{A_{re,sur}} u_{A_{re,sur}2} - \frac{\partial g(X_2)}{\partial \lambda_{tube}}\sigma_{\lambda_{tube}} u_{\lambda_{tube}2}}{\sqrt{\left(\frac{\partial g(X_2)}{\partial T_{re,sur}}\sigma_{T_{re,sur}}\right)^2 + \left(\frac{\partial g(X_2)}{\partial \dot{m}_{st}}\sigma_{\dot{m}_{st}}\right)^2 + \left(\frac{\partial g(X_2)}{\partial A_{re,sur}}\sigma_{A_{re,sur}}\right)^2 + \left(\frac{\partial g(X_2)}{\partial \lambda_{tube}}\sigma_{\lambda_{tube}}\right)^2}} =$$

= -7.8661

$$\alpha_{T_{re,sur}2} = \frac{\left.\frac{\partial g(X_2)}{\partial T_{re,sur}}\right|_{\mu_{T_{re,sur}}}\sigma_{T_{re,sur}}}{\sqrt{\left(\frac{\partial g(X_2)}{\partial T_{re,sur}}\sigma_{T_{re,sur}}\right)^2 + \left(\frac{\partial g(X_2)}{\partial \dot{m}_{st}}\sigma_{\dot{m}_{st}}\right)^2 + \left(\frac{\partial g(X_2)}{\partial A_{re,sur}}\sigma_{A_{re,sur}}\right)^2 + \left(\frac{\partial g(X_2)}{\partial \lambda_{tube}}\sigma_{\lambda_{tube}}\right)^2}} = -0.0181$$

$$\alpha_{\dot{m}_{st}2} = -\frac{\left.\frac{\partial g(X_2)}{\partial \dot{m}_{st}}\right|_{\mu_{\dot{m}st}}\sigma_{\dot{m}_{st}}}{\sqrt{\left(\frac{\partial g(X_2)}{\partial T_{re,sur}}\sigma_{T_{re,sur}}\right)^2 + \left(\frac{\partial g(X_2)}{\partial \dot{m}_{st}}\sigma_{\dot{m}_{st}}\right)^2 + \left(\frac{\partial g(X_2)}{\partial A_{re,sur}}\sigma_{A_{re,sur}}\right)^2 + \left(\frac{\partial g(X_2)}{\partial \lambda_{tube}}\sigma_{\lambda_{tube}}\right)^2}} = 0.8627$$

$$\alpha_{A_{re,sur}2} = -\frac{\left.\frac{\partial g(X_2)}{\partial A_{re,sur}}\right|_{\mu_{A_{re,sur}}}\sigma_{A_{re,sur}}}{\sqrt{\left(\frac{\partial g(X_2)}{\partial T_{re,sur}}\sigma_{T_{re,sur}}\right)^2 + \left(\frac{\partial g(X_2)}{\partial \dot{m}_{st}}\sigma_{\dot{m}_{st}}\right)^2 + \left(\frac{\partial g(X_2)}{\partial A_{re,sur}}\sigma_{A_{re,sur}}\right)^2 + \left(\frac{\partial g(X_2)}{\partial \lambda_{tube}}\sigma_{\lambda_{tube}}\right)^2}} = -0.0630$$

$$\alpha_{\lambda_{tube}2} = -\frac{\left.\frac{\partial g(X_2)}{\partial \lambda_{tube}}\right|_{\mu_{\lambda_{tube}}}\sigma_{\lambda_{tube}}}{\sqrt{\left(\frac{\partial g(X_2)}{\partial T_{re,sur}}\sigma_{T_{re,sur}}\right)^2 + \left(\frac{\partial g(X_2)}{\partial \dot{m}_{st}}\sigma_{\dot{m}_{st}}\right)^2 + \left(\frac{\partial g(X_2)}{\partial A_{re,sur}}\sigma_{A_{re,sur}}\right)^2 + \left(\frac{\partial g(X_2)}{\partial \lambda_{tube}}\sigma_{\lambda_{tube}}\right)^2}} = -0.5015$$

3) Compute the coordinates of the new design point using Equation 59 and 60:

The coordinates of the new design point in the original space using Equation 60:

$$T_{re,sur3} = \mu_{T_{re,sur}} + \beta_2 \sigma_{T_{re,sur}} \alpha_{T_{re,sur}2} = 608.5481$$

$$\dot{m}_{st3} = \mu_{\dot{m}_{st}} + \beta_2 \sigma_{\dot{m}_{st}} \alpha_{\dot{m}_{st}2} = 0.9642$$

$$A_{re,sur3} = \mu_{A_{re,sur}} + \beta_2 \sigma_{A_{re,sur}} \alpha_{A_{re,sur}2} = 1.0743$$

$$\lambda_{tube3} = \mu_{\lambda_{tube}} + \beta_2 \sigma_{\lambda_{tube}} \alpha_{\lambda_{tube}2} = 28.6140$$

The coordinates of the new design point in the normal standard space using Equation 59:

$$u_{T_{re,sur}3} = \frac{T_{re,sur3} - \mu_{T_{re,sur}}}{\sigma_{T_{re,sur}}} = 0.1425$$

$$u_{\dot{m}_{st}3} = \frac{\dot{m}_{st3} - \mu_{\dot{m}_{st}}}{\sigma_{\dot{m}_{st}}} = \text{-}6.7860$$

$$u_{A_{re,sur}3} = \frac{A_{re,sur3} - \mu_{A_{re,sur}}}{\sigma_{A_{re,sur}}} = 0.4953$$

$$u_{\lambda_{tube}3} = \frac{\lambda_{tube3} - \mu_{\lambda_{tube}}}{\sigma_{\lambda_{tube}}} = 3.9447$$

The probability of failure for β_2 using Equation 49 is:

$$P_f = \Phi(-\beta_2) = 1.0000$$

4) Check β convergence:

$$\varepsilon = \left| \frac{\beta_2 - \beta_1}{\beta_1} \right| = 4.3580$$

Since $\varepsilon \geq \varepsilon_r$, ($\varepsilon_r = 0.0001$) continue the process.

4.5.2. Simulation Results

The same procedure, as in iteration (2), can be repeated until the stopping criterion ($\varepsilon > \varepsilon_r$) is satisfied. The iteration results are summarized in Table 4. The safety-index β is 2.4731.

Table 4. HL-RF iteration results for STPP example

Iteration number	1	2	3	4	5	6	7
$g(X_k)\ \times 10^7$	0.5106	-0.7429	1.3196	0.0389	0.0004	0.0000	0.0000
$\nabla g(T_{re,sur}) \times 10^4$	1.1938	0.0250	1.4986	0.8097	0.7869	0.7854	0.7852
$\nabla g(\dot{m}_{st}) \times 10^6$	-2.3415	-2.3782	-1.8526	-2.3574	-2.3591	-2.3590	-2.3590
$\nabla g(A_{re,sur}) \times 10^7$	1.2236	0.0347	1.4418	1.1306	1.1501	1.1527	1.1529
$\nabla g(\lambda_{tube}) \times 10^5$	5.0736	3.4706	5.2361	3.3730	3.3113	3.3074	3.3068
β	2.3425	-7.8661	2.3147	2.4753	2.4731	2.4731	2.4731
$\alpha_{T_{re,sur}}$	- 0.3289	-0.0181	-0.3615	-0.2500	-0.2404	-0.2396	-0.2395
$\alpha_{\dot{m}_{st}}$	0.3225	0.8627	0.2235	0.3640	0.3604	0.3598	0.3598
$\alpha_{A_{re,sur}}$	- 0.8428	-0.0630	-0.8695	-0.8729	-0.8785	-0.8791	-0.8791
$\alpha_{\lambda_{tube}}$	- 0.2784	-0.5015	-0.2516	-0.2075	-0.2015	-0.2009	-0.2009
$T_{re,sur2}$ °C	553.7743	608.548	549.790	562.863	564.325	564.448	564.461
$\dot{m}_{st2}$ kg/s	3.2267	0.9642	3.1552	3.2703	3.2674	3.2669	3.2669
$A_{re,sur2}$ m^2	0.7039	1.0743	0.6981	0.6759	0.6741	0.6739	0.6739
λ_{tube2} W/m K	23.1207	28.6140	23.2041	23.2863	23.3045	23.3061	23.3063

Iteration number	1	2	3	4	5	6	7
$u_{T_{re,sur}}2$	- 0.7704	0.1425	-0.8368	-0.6189	-0.5946	-0.5925	-0.5923
$u_{\dot{m}_{st}}2$	0.7556	-6.7860	0.5172	0.9011	0.8913	0.8898	0.8897
$u_{A_{re,sur}}2$	-1.9742	0.4953	-2.0127	-2.1607	-2.1725	-2.1740	-2.1741
$u_{\lambda_{tube}}2$	- 0.6522	3.9447	-0.5823	-0.5136	-0.4983	-0.4969	-0.4968
ε	-	4.3580	1.2943	0.0694	0.0009	0.0000	0.0000

Since the limit-state function value at MPP ($T_{re,sur}$ = 564.461, $\dot{m}_{st}$ = 3.2669, $A_{re,sur}$ = 0.6739, λ_{tube} = 23.3063) is close to zero compared to the starting value, this safety-index can be considered as the shortest distance from the origin to the limit-state surface. The new design point at MPP is the optimal design parameters ensuring the objective function (F_{up} =10 %).

In order to investigate the variation of the probability of failure, we must vary the unsatisfactory performance factor (F_{up}) from 0.1 to 0.9 following the same HL-RF algorithm procedure mentioned previously.

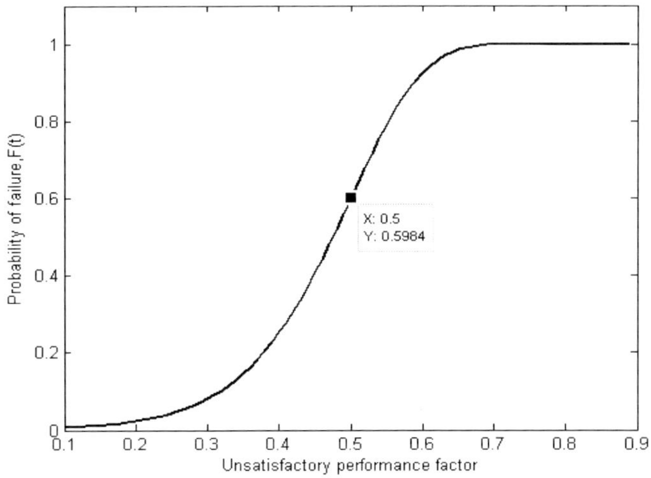

Figure 14. Failure function of the STPP unsatisfactory performance factor for $A_{re,sur}$ = 1 m^2.

Figure 14 shows the probability of failure function or cumulative distribution function (CDF) of the unsatisfactory performance factor (F_{up}). This plot can be used to predict the probability of F_{up} being less (or more) than a particular value, or between two values. For example, in this graph, there is approximately a 60 % probability that F_{up} will be less than 50 % and 40% probability that F_{up} will be greater than 50 %. There is approximately a 0.6 – 0.1= 0.5 (50%) probability that F_{up} will be between 30 % and 50 %. In other words, the probability of risk (40%) is too high so it is important to decrease this probability. In this case we must know the design parameter influencing on this probability.

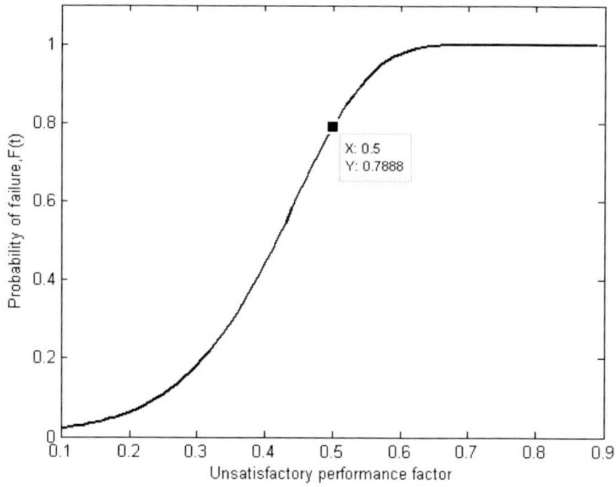

Figure 15. Failure function of the STPP unsatisfactory performance factor for $A_{re,sur} = 0.9\ m^2$.

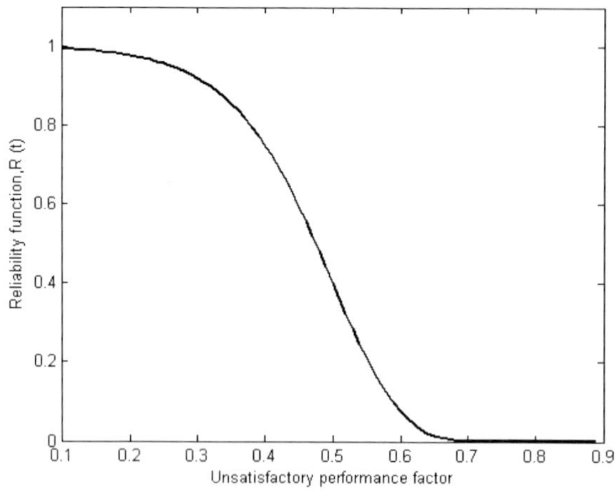

Figure 16. Reliability function of the STPP unsatisfactory performance factor for $A_{re,sur} = 1\ m^2$.

To this end it is necessary to analyze the sensitivity factor α_i, the parameter representing the high sensitivity factor is the parameter which has a great influence. In this example the receiver surface area $A_{re,sur}$ is the parameter having the highest sensitivity factor $\left(\alpha_{A_{re,sur}}\right)^2 = 0.7728$.

The Figure 15 another simulation example where $A_{re,sur}$ = 0.9 m^2 in the initial design point. We notice that, there is approximately 80 % probability that F_{up} will be less than 50 % and 20% probability that F_{up} will be greater than 50 %. So in this case the probability of risk is decreased by 20%.

The graph of Figure 16 shows the variation of the reliability function (R = 1-F) with the variation of unsatisfactory performance factor F_{up}. When F_{up} increases, the reliability of STPP decreases (when $F_{up} \leq 10\%$, R = 1 and when $F_{up} \geq 70\%$, R = 0). Physically, the reliability of any system decrease when there is loss of performances.

CONCLUSION

This chapter presented the fundamentals and approaches of performance and reliability analysis applied to solar tower power plants (STPP). General description of STPP with storage system has been carried out in order to show its functioning mode and its main subsystems. Artificial neural network (ANN) method has been used to analyze the STPP performance. ANN analysis and methodology have been described. In order to show the efficacy of this method, an example of hypothetical STPP has been introduced. A good convergence and accuracy between calculated and observed values has been noted.

An overview of main reliability analysis methods and tools has been presented. First order reliability method (FORM) has been chosen in order to study the reliability of the hypothetical STPP. The reliability analysis seems suitable to estimate the safety index of the receiver STPP and can be used as a guide to identify the most probable point (MPP). So, FORM has the ability to correct the first proposed design, then to give new safe design.

References

Akbari, M., Khazaee, P., Sabetghadam, I., and Karimifard, P. Failure modes and effects analysis (FMEA) for power transformers. *28th International Power System Conference* - 2013 Tehran, Iran.

Al-Shamisi, M. H., Assi, A. H. and N-Hejase, H. A. *Using MATLAB to develop artificial neural network models for predicting global solar radiation in Al Ain City – UAE*, Ali H. Assi (Ed.), ISBN 978-953-307-656-0, InTech (2011).

Ayyub, B. and Mccuen, R. *Probability, Statistics and Reliability for engineers*. CRC Press New York (1997).

Becker, M. and Klimas, P. C. *Second-Generation Central Receiver Technologies: A Status Report,* eds. Muller CF, Karlsruhe, Germany (1993).

Benammar, S. and Benazzouz, D. Fault Detection and Isolation using Levenberg-Marquardt algorithm, *6th International Symposium on Hydrocarbons and Chemistry Boumerdes*, Algeria, 13-15 October 2012.

Benammar, S., Khellaf, A. and Mohammedi, K. Contribution to the modeling and simulation of solar power tower plants using energy analysis. *Energy Conversion and Management*, 2014, 78, 923–930.

Benazzouz, D., Benammar, S. and Adjerid, S. Fault Detection and Isolation Based on Neural Networks Case Study: Steam Turbine. *Energy and Power Engineering*, 2011, 3, 513-516.

Blair, N., Mehos, M. and Christensen, C. Modeling photovoltaic and concentrating solar power trough performance, cost and financing with the Solar Advisor Model. *NREL CP-670-42922,* pp. 1–7 (2008a).

Blair, N., Mehos, M. and Christensen, C. Sensitivity of concentrating solar power trough performance, cost and financing with the Solar Advisor Model. *NREL CD-550-42702*, pp. 1–8 (2008b).

Blair, N., Dobos, A P. and Freeman, J. Ty Neises, and Michael Wagner. System Advisor Model, SAM 2014.1.14: General description. *Technical Report*, NREL/TP-6A20-61019 (2014).

Birolini, A. *Quality and reliability of technical systems*. Ed. Springer (1997).

Caner, M., Gedik, E. and Kecebas, A. Investigation on thermal performance calculation of two type solar air collectors using artificial neural network, *Exp. Syst. Appl.*, 2011, 38, 1668–1674.

Chellali, F., Khellaf, A. and Belouchrani, A. Wavelet spectral analysis of the temperature and wind speed data at Adrar, Algeria. *Renewable Energy*, 2010, 35 (6), 1214-1219.

Chellali, F., Khellaf, A., Belouchrani, A., and Khanniche, R. A comparison between wind speed distributions derived from the maximum entropy principle and Weibull distribution. Case of study; six regions of Algeria. *Renewable and Sustainable Energy Reviews*, 2012, 16 (1), 379-385.

Choi, S. K., Grandhi, R. V. and Canfield, R. A. *Reliability-based structural design*, Springer-Verlag, London (2007).

Collins, E., Dvorack, M., Mahn, J., Mundt, M., and Quintana, M. In: *Reliability and availa-bility analysis of a fielded photovoltaic system.* Photovoltaic specialists conference (PVSC), 2009 34th IEEE; 2009. p. 002316–21.

David, J. S. *Reliability, Maintainability and Risk: Practical Methods for Engineers*. Elsevier (2011).

Dersch, J., Hennecke, K. and Quaschning, V. Greenius – a simulation tool for renewable energy utilisation. In: *Eurosun 2008 Conference*, Lisbon, Portugal (2008).

Do, M. T., Sprooten, J., Clenet, S., and Robyns, B. "Reliability evaluation of power system with large-scale wind farm integration using first-order reliability method", *14th European Conference on Power Electronics and Applications*, pp. 1-10, 2011.

Ehnberg, J. Generation Reliability for Isolated Power Systems with Solar, Wind and Hydro Generation. *Technical Report No. 478L ISSN: 1651-4998* Department of Electric Power Engineering Chalmers University of Technology SE-412 96 GÄoteborg, Sweden.

Esen, H., Inalli, M., Sengur, A., and Esen, M. Forecasting of a ground-coupled heat pump performance using neural networks with statistical data weighting preprocessing, *Int. J. Therm. Sci.*, 2008a, 47 (4), 431–441.

Esen, H., Inalli, M., Sengur, A., and Esen, M. Performance prediction of a ground coupled heat pump system using artificial neural networks, *Exp. Syst. Appl.*, 2008b, 35 (4), 1940–1948.

Falcone, P. K. *A handbook for solar central receiver design*. SAND 86-8009, Sandia National Laboratories Livermore (1986).

García, I. L., Álvarez, J. L. and Blanco, D. Performance model for parabolic trough solar thermal power plants with thermal storage: Comparison to operating plant data. *Solar Energy,* 2011, 85, 2443–2460.

Garrol, A. and Barrara, F. Reliability Analysis of Residential Photovoltaic Systems. *Inter-national Conference on Renewable Energies and Power Quality (ICREPQ'11),* Las Palmas de Gran Canaria (Spain), 13th to 15th April, 2011.

Hennecke, K., Dersch, J. and Quaschning, V. Greenius – a simulation tool for renewable energy utilization. In: *SolarPACES 2010 Conference*, Perpignan, France (2010).

Ho, C. K. Software and codes for analysis of concentrating solar power technologies. *Sandia report,* SAND2008-8053 (2008).

Ho, C. K. and Kolb, G. J. Incorporating Uncertainty into Probabilistic Performance Models of Concentrating Solar Power Plants. *Journal of Solar Energy Engineering*, 2010, Vol. 132 / 031012.

Hoang, P. *Handbook of reliability engineering*. Ed Springer (2003).

Leemis, M. L. Probabilistic models and statistical methods in reliability. In: *IEEE Pro-ceedings Annual Reliability and Maintainability Symposium, Tutorial notes*, US (1994).

Li, X., Kong, W., Wang, Z., Chang, C., and Bai, F. Thermal model and thermodynamic performance of molten salt cavity receiver. *Renewable Energy*, 2010, 35, 981–988.

Lippke, F. Simulation of the Part-Load Behavior of a 30 MWe SEGS Plant, *SANDIA Report* SAND95-1293 (1995).

Lu, J. and Ding, J. *Heat Transfer Performances and Exergetic Optimization for Solar Heat Receiver, Evaporation, Condensation and Heat transfer*, Dr. Amimul Ahsan (Ed.), ISBN: 978-953-307-583-9, InTech (2011).

Modi, A. and Haglind, F. Performance analysis of a Kalina cycle for a central receiver solar thermal power plant with direct steam generation. *Applied Thermal Engineering*, 2014, 65 201-208.

Moharil, R. M. and Kulkarni, P. S. Reliability analysis of solar photovoltaic system using hourly mean solar radiation data. *Solar Energy*, 2010, 84, 691–702.

Montes, M. J., Abánades, A. and Martínez-Val, J. M. Performance of a direct steam generation solar thermal power plant for electricity production as a function of the solar multiple. *Solar Energy*, 2009, 83, 679–689.

Nithyanandam, K. and Pitchumani, R. Cost and performance analysis of concentrating solar power systems with integrated latent thermal energy storage, *Energy,* 2014, 64, 793-810.

James, V. J. *Integrated Logistics Support Handbook.* McGraw–Hill Professional, 3rd edition (2006), ISBN 0-07-147168-5.

Jones, S. A., Pitz-Paal, R., Schwarzboezl, P., Blair, N., and Cable, R. TRNSYS modeling of the SEGS VI parabolic trough Solar Electric Generating System. In: *Proceedings of ASME International Solar Energy Conference Solar Forum (2001).*

Kalogirou, S. A. Artificial neural networks in renewable energy systems applications: a review, Renew. *Sustainable Energy Rev.*, 2001, 5, 373–401.

Kalogirou, S. A. Prediction of flat-plate collector performance parameters using artificial neural networks, *Sol. Energy*, 2006, 80, 248–259.

Keshtegar, B. and Miri, M. An Enhanced HL-RF Method for the Computation of Structural Failure Probability Based On Relaxed Approach. *Civil Engineering Infrastructures Journal*, 2013, 46 (1): 69 – 80, ISSN: 2322 – 2093.

Kishore, L. N. and Fernandez, E. In: *Reliability well-being assessment of PV-wind hybrid system using monte carlo simulation.* Emerging trends in electrical and computer techno-logy (ICETECT), 2011 International conference on; 2011. p. 63–8.

Kurt, H., Atik, K., Ozkaymak, M., and Recebli, Z. Thermal performance parameters estima-tion of hot box type solar cooker by using artificial neural network, *Int. J. Therm. Sci.*, 2008, 47, 192–200.

Kistler, B. L. A user's manual for DELSOL3: A computer code for calculating the optical performance and optimal system design for solar thermal central receiver plants. *Sandia report,* SAND 86-8018 (1986).

Paliwal, P., Patidar, N. P. and Nema, R. K. A novel method for reliability assessment of autonomous PV-wind-storage system using probabilistic storage model. *Electrical Power and Energy Systems*, 2014, 55, 692–703.

Patnode, A. M. Simulation and performance evaluation of parabolic trough solar power plants. *Ph.D. thesis (2006),* University of Wisconsin-Madison. Department of Mechanical Engineering.

Price, H., Svoboda, P. and Kearney, D. Validation of the FLAGSOL parabolic trough solar power plant performance model. *NREL TP-471-7297*, pp. 1–6 (1995).

Price, H. A parabolic trough solar power plant simulation model. *NREL CP-550-33209,* pp. 1–9 (2003).

Rackwitz, R. and Fiessler, B. Structural reliability under combined load sequences. *Computers and Structures*, 1978, 9(8), 489–494.

Saheb-Koussa, D., Haddadi, M. and Belhamel, M. Economic and technical study of a hybrid system (wind–photovoltaic–diesel) for rural electrification in algeria. *Appl. Energy*, 2009, 86, (7–8):1024–30.

Santosh, T. V., Saraf, R. K., Ghosh, A. K., and Kushwaha, H. S. Optimum step length selection rule in modified HL-RF method for structural reliability. *International journal of Pressure Vessels Piping*, 2006, 83(10), 742–748.

Schwarzbözl, P., Buck, R., Sugarmen, C., Ring, A., Crespo, J. M., Altwegg, P., and Enrile, J. Solar gas turbine systems: Design, cost and perspectives. *Solar Energy*, 2006, 80, 1231–1240.

Saheb-Koussa, D., Haddadi, M. and Belhamel, M. Economic and technical study of a hybrid system (wind–photovoltaic–diesel) for rural electrification in algeria. *Appl. Energy*, 2009, 86, (7–8):1024–30.

Stoddard, M. C., Faas, S. E., Chiang, C. J., Dirks, J. A. SOLERGY – A Computer Code for Calculating the Annual Energy from Central Receiver Power Plants. *SANDIA Report* SAND86-8060 (1987).

Telsnig, T., Eltrop, L. and Winkler, H. Efficiency and Costs of Different Solar Power Plant Configuration for Sites in Gauteng and the Northern Cape; SA. *1st Southern African Solar Energy Conference,* 21 - 23 May 2012.

Tymvios, F., Michaelides, S. and Skouteli, C. Estimation of Surface solar radiation with Artificial neural networks, In: *Modeling Solar Radiation at the Earth Surface,Viorel Badescu,* pp. (221-256), Springer, ISBN 978-3-540-77454-9, Germany (2008).

Urbina, A., Paez, T. L., O'Gorman, C. C., and Barney, P. *Reliability of rechargeable batteries in a photovoltaic power supply system.* Sandia National Laboratories, SAN 098-2658C (1998).

Villemeur, A. *Sûreté de fonctionnement des systèmes industriels: Fiabilité, Facteurs humains et Informatisation*. Collection EdF - Ed Eyrolles (1988).

Wagner, M. J., Blair, N. and Dobos, A. A detailed physical trough model for NREL's Solar Advisor Model. *NREL CP-5500-49368*, pp. 1–8 (2010).

Wu, J., Yan, S. and Xie, L. Reliability analysis method of a solar array by using fault tree analysis and fuzzy reasoning Petri net. *Acta Astronautica*, 2011, 69, 960–968.

Xu, C., Wang, Z., Li, X., Sun, F. Energy and exergy analysis of solar power tower plants. *Applied Thermal Engineering*, 2011, 31, 3904-3913.

Xu, E., Yu, Q., Wang, Z., and Yang, C. Modeling and simulation of 1 MW DAHAN solar thermal power tower plant. *Renewable Energy,* 2011, 36, 848-857.

Yu, Q., Zhifeng, W. and Xu, E. Simulation and analysis of the central cavity receiver's performance of solar thermal power tower plant. *Solar Energy*, 2012, 86, 164–174.

Zhang, Q., Li, X., Chang, C., Wang, Z., and Liu, H. An experimental study: Thermal performance of molten salt cavity receivers. *Applied Thermal Engineering* 50, 2013, 334-341.

Zini, G., Mangeant, C. and Merten J. Reliability of large-scale grid-connected photovoltaic systems. *Renewable Energy*, 2011, 36(9): 2334–2340.

In: Solar Power
Editor: Stephen Bailey

ISBN: 978-1-63321-317-3

Chapter 5

LIQUID CRYSTAL DEVICES FOR PHOTOVOLTAIC SYSTEMS

Yi-Hsin Lin [*] ***and Yu-Shih Tsou***
Department of Photonics, National Chiao Tung University

ABSTRACT

The power conversion efficiency, which is defined as the ratio of the output electrical power density to the incident light power density of the solar cell, is one of the most important parameters of a photovoltaic (PV) system. The efficiency depends on the photon flux density of the incident light, which changes frequently in different environment. For example, the solar cell in a solar power station receives the sunlight, which changes with the weather, season, and time; the solar cell on a mobile device may be used indoor and outdoor alternately. In order to preserve a static output of a solar cell in different environments, a tunable light modulator for keeping a static incident photon flux density is needed. Liquid crystal (LC) light modulators have been developed for years, which can be used to improve the PV systems. In this chapter, we introduce the PV systems with high and steady output by adopting electrically tunable LC light modulators. This concept may open a window to enhance the efficiency of CPV systems by using active optical elements.

[*] Corresponding author's email:yilin@mail.nctu.edu.tw.

1.1. INTRODUCTION

Solar photovoltaic (PV) energy conversion is a one-step conversion process that generates electrical energy from light energy [1–6]. The PV effect was first reported by the French physicist Alexandre-Edmund Becquerel in 1839 when he observed that the action of light on a silver-coated platinum electrode immersed in an electrolyte solution produced an electric current. In recent years, using renewable energy sources, such as sunlight, wind, rain, tides, waves, and geothermal heat, has been becoming critical because of the global energy crisis. The use of solar PV energy is considered a potential solution to the energy problem because an almost infinite amount of energy can be harnessed from solar radiation, and is sufficient to power industries worldwide. However, using solar PV energy is still not prevalent because it is not sufficiently efficient. Therefore, enhancing the efficiency of solar PV energy devices is imperative.

The efficiency of solar PV energy devices depends on the materials and the photon flux density of the incident light. Multiple methods for managing light have been developed, including antireflection coatings, concentrator systems, solar trackers, light trapping structures, and photon recycling systems. Antireflection coatings help to reduce the reflectance at the surface of the solar cells. Concentrator can concentrate a large amount of sunlight onto a small area of solar cells, which needs a solar tracker to keep the incident light at the normal direction. Light trapping structures increase the path length of photons inside the solar cell, which results in the increasing of the probability of photogeneration per incident photon. Photon recycling systems help to increase the percentage of the absorbed energy. Although these methods can increase the efficiency of the PV systems, they are still unable to function at an optimal status consistently because weather and seasonal changes affect ambient light, which diminishes the amount of energy generated from a solar cell. Because we cannot change the physical limitations of the materials used in an existing PV system, using an additional device that enables solar cells to constantly function optimally is reasonable. Using LC devices is a potential solution because of LC's electrooptical properties. Some groups have already attempted to combine LC devices with PV systems. The anisotropic and ordering properties of LCs have been used to improve the efficiency of a dye-sensitized solar cell, and the thermal properties of cholesteric LCs have been used to determine local shunts. A polymer solar cell can also be used as a polarizer in LCDs to recycle the energy of the absorbed light. In this study, we adopted an LC light modulator, which is a combination of concentrator, sun tracker, and other techniques, into an existing PV system to enable it to function efficiently under various ambient light conditions. Our research goal was to realize a PV system with a stable maximal electric output by adopting an electrically tunable LC light modulator.

1.2. Theory

1.2.1. Photogeneration

When the photons from the sun are incident to a solar cell, the photons are absorbed in the depletion region. The electron-hole pairs in the depletion region are excited when the energy of the photons is larger than that of the diode band gap. The electrons are then excited from the valence band to the conduction band, and generate the free electrons and holes. The free electrons and the holes move toward the electrodes to form a photocurrent. The electric current (I) or DC resulting from the drifting electrons and holes can be used as an electric power source when wires are used to connect the solar cell to the loads. The DC can also be converted to AC by using a DC-AC inverter [7-18].

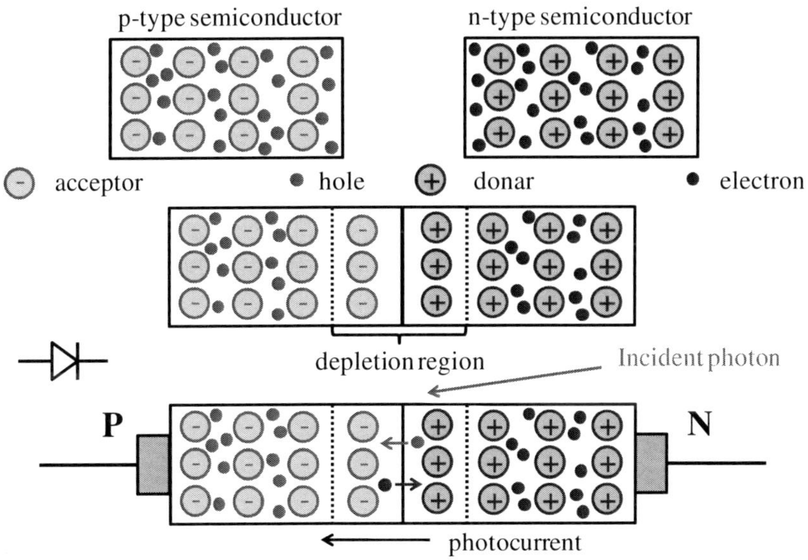

Figure 1-1. Photogeneration in a semiconductor solar cell.

To discuss the effects of the concentration ratio on photogeneration, we first assume that the solar cell operates in low injection condition, which means the photogenerated carrier density is smaller than the equilibrium majority carrier density. In low injection conditions, a concentrating photovoltaic (CPV) cell can be essentially treated using the standard theory for a p-n junction. In low injection conditions, the minority carrier recombination rates are linear and the current densities J resolve into a dark component J_{dark} and a light-generated component J_{SC}.:

$$J = J_{SC} - J_0(e^{qV/mk_BT} - 1) \tag{1-1}$$

where m is an ideality factor and the dark current typically has a diode-like form.

Now consider that the incident flux b_s is increased by a concentration ratio X through the concentration of the light, J_{SC} increases to a first approximation by the same factor, such as $J_{SC}(Xb_s) \approx X \cdot J_{SC}(b_s)$, but leave the dark current, for a given bias, unaffected. Therefore, the net current increases with X whenever the given bias is below the V_{OC}. Moreover, a higher bias is required to achieve the open circuit condition where the dark current cancels the photocurrent. By replacing J_{SC} with $X \cdot J_{SC}$ in Eq. (1-1), the open circuit for an ideal diode becomes

$$V_{OC}(X) = \frac{m \cdot k_B \cdot T}{q} \ln\left(\frac{X \cdot J_{SC}}{J_0} + 1\right) \approx V_{OC}(1) + \frac{m \cdot k_B \cdot T}{q} \ln X \tag{1-2}$$

where $V_{OC}(1)$ represents the open circuit voltage under unconcentrated sunlight (1 sun). According to Eq. (1-2), the V_{OC} increases logarithmically with X. In the ideal case that the fill factor remains the same and the series resistances are neglected, the power delivered by the solar cell should increase by a factor of

$$X\left(1 + \frac{m \cdot k_B \cdot T}{q \cdot V_{OC}(1)} \ln X\right)$$

and the efficiency also increases by a factor of

$$\left(1+\frac{m\cdot k_B\cdot T}{q\cdot V_{OC}(1)}\ln X\right).$$

In high injection conditions, the photogenerated carrier density is larger than the doping density in the base, and both carrier types must be included in the transport equations. At injection levels where radiative and Auger processes are dominant, the superposition of the light-dependent term and the bias-dependent term becomes invalid for the recombination-generation current. The dark-current term increases with generation rate and the V_{OC} rises more slowly than predicted in Eq. (1-2).

Up to now we have derived the relation between the output power, efficiency, and the concentration ratio in an ideal case. However, the series-resistance effect cannot be completely eliminated in all types of solar cell. By increasing the current density, the concentration increases the potential drop because of the series resistance in solar cells. This is a problem particularly in the emitter layer where lateral current densities are high. Emitter doping levels are limited by recombination and cannot be increased indefinitely, nor can the contacted area become too large; thus, a compromise or an optimization method must be devised.

The series resistance arises from the resistance of the cell material to the current flow, particularly through the front surface to the contacts, and from resistive contacts. Assume the equivalent series resistance is R_S for the entire solar cell and the power dissipation on the equivalent series resistance can be expressed as $J_{SC}^{\ 2}\times R_S$. The output electrical power density P_{out} depending on X can be expressed as

$$P_{out}(X)=X\times P_1\times\{1+(k_B\times T/q\times V_{OC,1})\times\ln(X)\}-\{X^2\times J_{SC,1}^{\ 2}\times R_S\} \qquad (1\text{-}3)$$

where P_1 is the ideal output power density and $J_{SC,1}$ denotes the short circuit current density under a 1-sun illumination. From Eq. (1-3), we determined that P_{out} is a function of the only variable X. Therefore, we can realize a CPV system with a stable electric output by adopting a constant X through the concentrator. To obtain the highest P_{out}, $P_{out}(X)$ should satisfy Eq. (1-4):

$$\frac{dP_{out}(X)}{dX}=0 \qquad (1\text{-}4)$$

According to Eq. (1-4), the output power density reaches a maximum when

$$k_B \times T/q \times FF \times \{\ln(X)+1\} - 2 \times X \times J_{SC,1} \times R_S + FF \times V_{OC,1} = 0 \quad (1\text{-}5)$$

is satisfied. The solution of X in Eq. (1-5) is a fixed number determined by the parameters of the materials and the structure of the solar cell. Therefore, for a specific solar cell, the maximum electric output power can be achieved by adopting an X that satisfies Eq. (1-5).

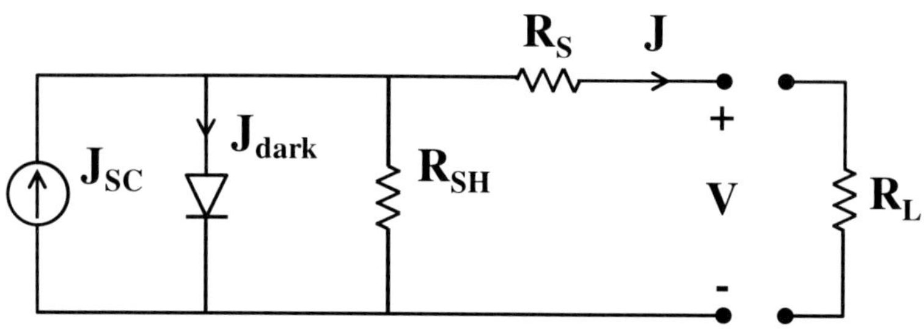

Figure 1-2. Equivalent circuit model.

The absorption spectrum of a solar cell depends on its materials. The spectrum of the sun's solar radiation is close to that of a black body with a temperature of approximately 5800 K. Most of the energy of the sunlight is contained within the wavelength from 300 nm to 1800 nm. However, the absorption spectrum of a single-junction silicon solar cell is from 400 to 1100 nm, which is much less than the spectrum of the sunlight.

The energy of the sunlight that has a wavelength larger than the band gap of the silicon solar cell (approximately 1100 nm or 1.1 eV) cannot be absorbed by the solar cell. To increase the absorption spectrum, multijunction solar cells were developed. Multijunction solar cells or tandem cells are solar cells containing several p-n junctions.

Each junction is tuned to a different wavelength of light, reducing one of the largest inherent sources of losses, and thereby increasing efficiency. The absorption spectrum is extended and closer to the spectrum of sunlight. Theoretically, a traditional single-junction cell has a maximum efficiency of 34%, whereas a theoretical infinite-junction cell would improve this to 87% under highly concentrated sunlight.

1.2.2. Optical Properties of Liquid Crystals

The optical properties of liquid crystals (LCs), such as optical rotation and birefringence, have been studied by various researchers and used in multiple applications [19–27]. Using LC devices is a potential solution for improving the performance of PV systems because these devices can modulate light. To study how LC devices modulate light, the physical properties of LCs are introduced in this section.

The continuum theory is often employed to investigate the distributions of LC molecules under external fields. The basic structure of continuum theory was first studied by Oseen and Zocher in 1933. Frank later critically reexamined the theory and presented the theory of curvature elasticity. LCs are either rod-like or disk-like molecules. From a macroscopic viewpoint, LC molecules tend to align in a common direction, which is defined as the *director*, n. The continuum theory describes the distortions of n under certain external forces. When external fields are applied, the LC molecules are rotated according to the field strength. When the field is removed, the LC molecules revert to their original alignments. This effect is similar to elastic deformation in solids. Based on the Hooke law, the elastic free energy density f_{LC} was studied by Frank:

$$f_{LC} = \frac{1}{2}\left[k_{11}(\nabla \cdot n)^2 + k_{22}(n \cdot \nabla \times n)^2 + k_{33}(n \times \nabla \times n)^2\right] \tag{1-6}$$

where k_{ii}, i = 1, 2, and 3 are the elastic constants associated with three basic types of LC deformation: splay, twist, and bend (Figure 1-3).

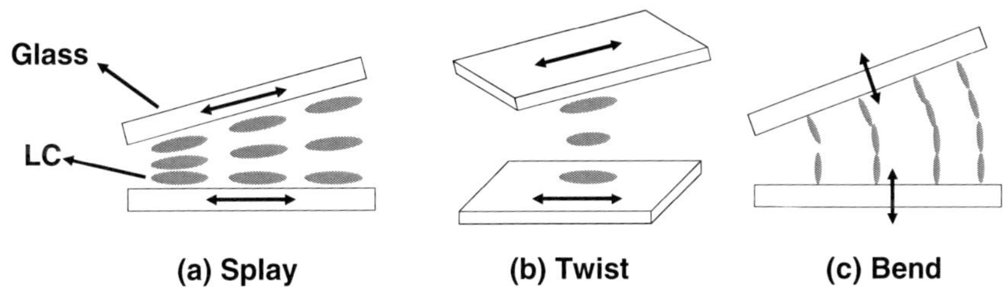

Figure 1-3. Schematic illustration of three deformations in LC medium.

When a monochromatic plane wave propagates in a homogeneous, nonabsorbing, and nonmagnetic material, the dispersion relation, which is a corollary from the Maxwell equation, must be satisfied:

$$\frac{k_x^{2}}{k^2 - k_0^{2} n_x^{2}} + \frac{k_y^{2}}{k^2 - k_0^{2} n_y^{2}} + \frac{k_z^{2}}{k^2 - k_0^{2} n_z^{2}} = 1 \tag{1-7}$$

where $k = \omega / c$ is the wave vector, and n_x, n_y, and n_z are the principal indices of refraction, defined by $n_i^{2} = \varepsilon_i$ (i = x, y, z) and $k_0^{2} = \omega^2 \mu_0 \varepsilon_0$. Eq. (1-7) can be rewritten as

$$\left(k_x^{2} + k_y^{2} + k_z^{2}\right)\left(\frac{k_x^{2} n_x^{2} + k_y^{2} n_y^{2} + k_z^{2} n_z^{2}}{n_x^{2} n_y^{2} n_z^{2}}\right) - k_0^{2}\left(\frac{k_y^{2} + k_z^{2}}{n_x^{2}} + \frac{k_x^{2} + k_z^{2}}{n_y^{2}} + \frac{k_x^{2} + k_y^{2}}{n_z^{2}}\right) - k_0^{4} = 0 \tag{1-8}$$

LC molecules are typically considered as a uniaxial material that has an optic axis parallel to the director. Assuming that the director is parallel to the z-axis, the principal indices of refraction are $n_x = n_y = n_o$ and $n_z = n_e$, where n_o and n_e are the ordinary and extraordinary refractive indices, respectively. Thus, Eq. (1-8) can be rewritten as

$$\left(\frac{k_x^{2} + k_y^{2} + k_z^{2}}{n_o^{2}} - k_0^{2}\right)\left(\frac{k_x^{2} + k_y^{2}}{n_e^{2}} + \frac{k_z^{2}}{n_o^{2}} - k_0^{2}\right) = 0 \tag{1-9}$$

The first and second terms in Eq. (1-9) represent the normal surface of the ordinary wave (O-wave) and the extraordinary wave (E-wave), respectively. For the O-wave, which means the polarization of the wave is perpendicular to the LC director, the refractive index equals n_o. For the E-wave, when the wave propagates on the x-z plane with the angle θ between the

optic axis (z-axis) and the direction of the propagation, the wave vector can be expressed as

$$\begin{aligned} k &= k_x \hat{x} + k_z \hat{z} \\ &= k_0 n_{eff}\left(\hat{x}\sin\theta + \hat{z}\cos\theta\right) \end{aligned} \tag{1-10}$$

where n_{eff} represents the effective refractive index of the E-wave. By substituting Eq. (1-10) into Eq. (1-9), we obtain the relation between the n_{eff} and θ:

$$\frac{1}{n_{eff}^{2}} = \frac{\sin^2\theta}{n_e^{2}} + \frac{\cos^2\theta}{n_o^{2}}. \tag{1-11}$$

Eq. (1-11) shows that n_{eff} is a function of θ, which means the refractive index of the LC is dependent on the orientation of the director. Therefore, the orientation of the LC directors can be controlled by applying electric fields to achieve an electrically tunable refractive index.

The refractive indices of LCs are also wavelength dependent. When a chromatic light wave propagates in the LC media, the refractive indices of LC can be considered using the Cauchy equation, which is the empirical relationship between the refractive index and wavelength of light for a particular transparent material. The most general form of the Cauchy equation can be expressed as

$$n(\lambda) = A + \frac{B}{\lambda^2} + \frac{C}{\lambda^4} + \cdots \tag{1-12}$$

where n is the refractive index, λ is the wavelength, and A, B, and C are coefficients, which can be determined by fitting the equation to measured refractive indices at known wavelengths in experiments. The refractive index n in Eq. (1-12) can be replaced by n_o, n_e, or n_{eff} for LCs. In our experiments, we measured the refractive indices of LCs at three different wavelengths and calculated the coefficients A, B, and C by fitting the Cauchy equation, thereby obtaining the refractive indices of LCs as functions of wavelength, which are useful for the discussion of broadband sunlight.

1.3. Fabrication and Measurements

1.3.1. Sample Preparation

1.3.1.1. Liquid Crystal Lens

To demonstrate the concentrating photovoltaic (CPV) system using liquid crystal (LC) lens, two identical LC lenses were prepared [28-47]. The structure of the LC lenses is shown in Figure 1-4. For each of the LC lenses, two 0.5-mm indium tin oxide (ITO) glass substrates coated with mechanically buffered poly(vinyl alcohol) layers were used. We stacked the two ITO glass substrates in antiparallel rubbing directions with a cell gap of 35 μm to align LC directors. One of the ITO layers was etched with a hole-pattern within a diameter of 3 mm to provide an inhomogeneous electric field to the LC directors. The empty cell was then filled with the nematic LC MLC-2070 (Merck, Δn = 0.26) and wires were attached to apply the voltages. Finally, the two LC lenses with orthogonal rubbing directions were stacked to obtain polarization-independent LC lenses.

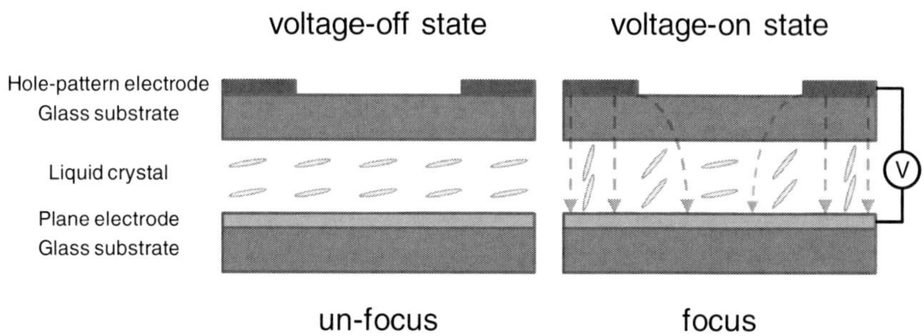

Figure 1-4. Structure of a hole-patterned LC lens.

1.3.1.2. Liquid Crystal and Polymer Composite film

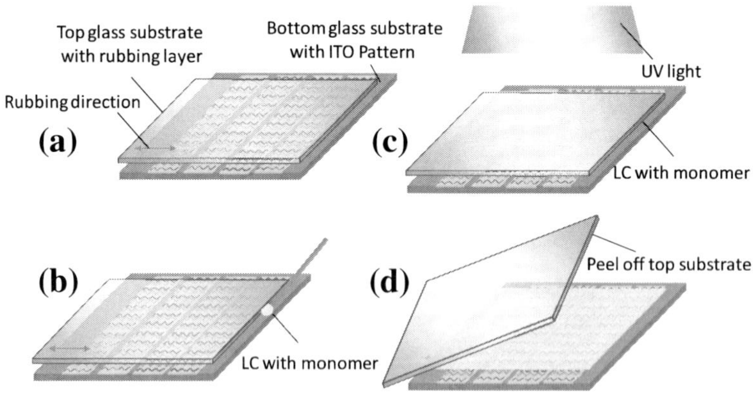

Figure 1-5. Fabrication process of liquid crystal and polymer composite film (LCPCF). (a) Prepare two glass substrates with a gap of ~7 μm. (b) Fill the mixture of LC and monomer into the cell. (c) Expose the cell to a UV light. (d) Peel of the top substrate.

An LC and polymer composite film (LCPCF) is a film that provides tunable surface free energy by controlling the orientation of the LC directors in the polymer grains [48–54]. Our group developed the LCPCF for years. We discussed the physical mechanisms of the surface and developed several applications based on droplet manipulation on an LCPCF, such as electrooptical switches, sperm testing devices, and liquid lenses. The droplet manipulation on an LCPCF can also be used in a CPV system as a concentrator and a sun tracker. Here we first describe the sample preparation of an LCPCF.

The fabrication process of the LCPCF is shown in Figure 1-5. To fabricate the LCPCF, we prepared a mixture containing a nematic LC mixture E7 (Merck) and a liquid crystalline monomer (4-(3-acryloyloxypropyloxy)-benzoic acid 2-methyl-1, 4-phenylene ester) at a 70:30 wt% ratio. The nematic LC (E7) consisted of four compounds: 5CB (4-pentyl-4'-cyanobiphenyl), 7CB (4-heptyl-4'-cyanobiphenyl), 8OCB (4-octyloxy-4'-cyanobiphenyl), and 5CT (4-pentyl-4'-cyanoterphenyl). The LC and monomer mixture was filled between two glass substrates with a gap of approximately 7 μm.

The top substrate was coated with a rubbed alignment layer. The bottom substrate was coated with pattered ITO electrodes. The ITO electrodes on the glass substrate were etched with interdigitated chevron patterns, shown as the zigzag electrodes.

The zigzag ITO strips have corner angles of 150°. The width and the gap of the electrode strips are 4 and 14 μm, respectively. The cell was then exposed to UV light with an intensity of 10 mW/cm^2 for approximately 60 min at a curing temperature of 70 °C. After phase separation and photopolymerization, the top glass substrate was peeled off by using a thermal-releasing process. A solidified LCPCF was then obtained. Figure 1-6 shows the scanning electron microscopy image of the LCPCF. The surface of the LCPCF comprises polymer grains and LCs are anchored among them. The average polymer grain size is approximately 70 nm, and the average domain size is approximately 225 nm.

Subsequently, a fluidic droplet was dropped on the LCPCF. The curvature of the droplet can be changed by applying a homogeneous electric field ($V_{total} = V_1 = V_2 = V_3$), and its position can be changed by applying a pulsed voltage to one of the electrode regions, as shown in Figure 1-7.

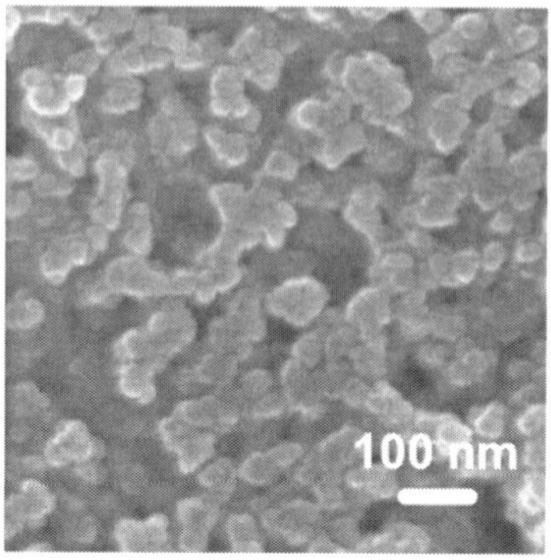

Figure 1-6. The scanning electron microscopy image of LCPCF [54].

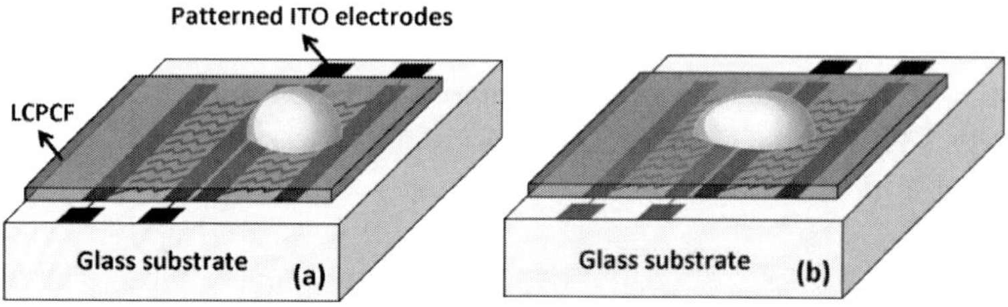

Figure 1-7. The droplet on LCPCF at (a) voltage off and (b) voltage on states.

1.3.2. Measurements

1.3.2.1. Focal Length

To measure the electrically tunable concentration ratios of the lenses, we first measured the electrically tunable focal lengths of the lenses. An unpolarized He-Ne laser (Melles Griot, 8144EU for $\lambda = 543.5$ nm) was used to demonstrate the concept. A crossed-polarizer method was used to measure the electrically tunable focal length of the LC lens. The incident light passed through a spatial filter and expander to obtain a collimated and uniform light beam. The LC lens was placed between crossed polarizers with the rubbing direction at 45° with respect to the axis of the polarizers. A relay lens was used to magnify the image. The transmittance distribution on the screen can then be obtained:

$$T = \frac{1}{2}\sin^2\frac{\Gamma}{2} \tag{1-13}$$

where the phase retardation Γ can be expressed as

$$\Gamma = \frac{2\pi}{\lambda}\left(n_{eff} - n_o\right)d \quad (1\text{-}14)$$

From Eq. (1-14), we determined that the difference of the phase retardation between every adjacent bright fringe or dark fringe is equal to 2 π. Thus, by counting the number of fringes (N) within the radius (r) of the LC lens, the focal length can be obtained using the following equation:

$$f = \frac{r^2}{2 \cdot \lambda \cdot N} \quad (1\text{-}15)$$

To measure the focal length of the liquid lens on the LCPCF, a homogeneous electric field ($V_{total} = V_1 = V_2 = V_3$) was applied to the LCPCF and the dynamics of the droplet were recorded using a charge-coupled device (CCD) camera (JAI CV-M30) with a frame rate of 120 frames/s. When voltages were applied to the LCPCF, the curvature radius of the droplet of the liquid lens was measured. According to the radius of curvature, we calculated the focal length of the liquid lens by using geometrical optics.

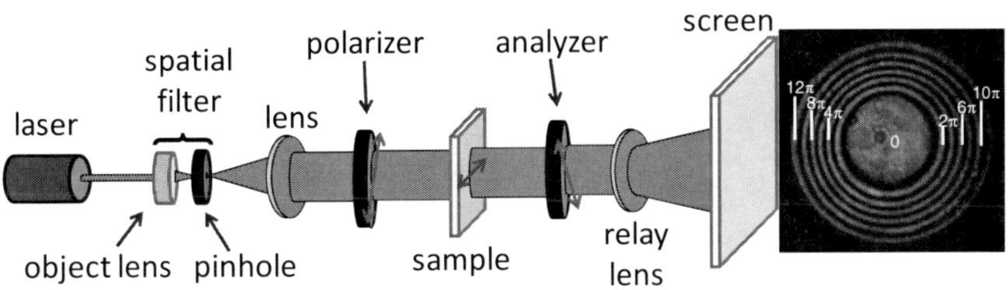

Figure 1-8. The experimental setup of focal length measurement.

1.3.2.2. Concentration Ratio

The concentration ratio is dependent on the aperture size of the concentrator, the area of the solar cell, focal length, and the distance between the concentrator and the solar cell. Figure 1-9 shows the experimental setup. Initially, the solar cell was set behind the lens at a distance equivalent to the minimal focal length of the lens, and the collimated light source was then applied. When the focal length of the lens was changed, the illuminated area at the plane where the solar cell was placed changed simultaneously. The illuminated area can be calculated using geometrical optics. The concentration ratio was then defined as the ratio of the aperture size of the lens to the illuminated area at the solar cell plane.

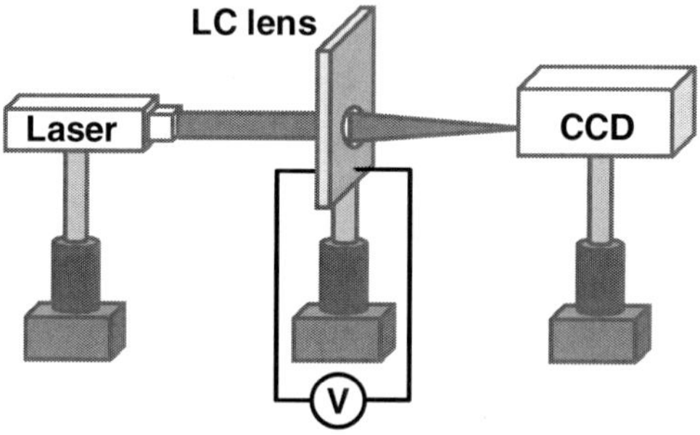

Figure 1-9. The concentration ratio can be controlled by the applied voltage of LC lens.

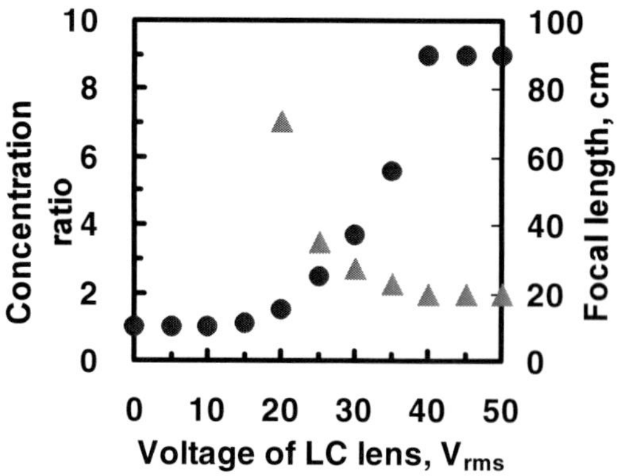

Figure 1-10. The concentration ratio (X) of the LC lens (blue dots) and the focal length (pink triangles) as a function of an applied voltage (V_{LC}) [29].

The concentration ratio X of the LC lens and the focal length as a function of the applied voltage V_{LC} are shown in Figure 1-10. In Figure 1-10, the concentration ratio increases from 1 to 9 with the applied voltage when the applied voltage is larger than the threshold voltage (approximately 15 V_{rms}). This is because the focal length decreased as the applied voltage decreased, and the illuminated area at the solar cell plane also decreased. The concentration ratio saturated at 9 when the V_{LC} exceeded 40 V_{rms}. This is because the focal length did not change with the applied voltage, as represented by $V_{LC} > 40\ V_{rms}$. Therefore, the maximum concentration ratio of the LC lens is 9.

1.3.2.3. Moving Distance

To measure the moving distance Δx of the liquid lens on the LCPCF induced by the inhomogeneous surface free energy, a pulsed voltage was applied to the LCPCF and the dynamics of the droplet were recorded using the same CCD camera, as shown in Figure 1-11.

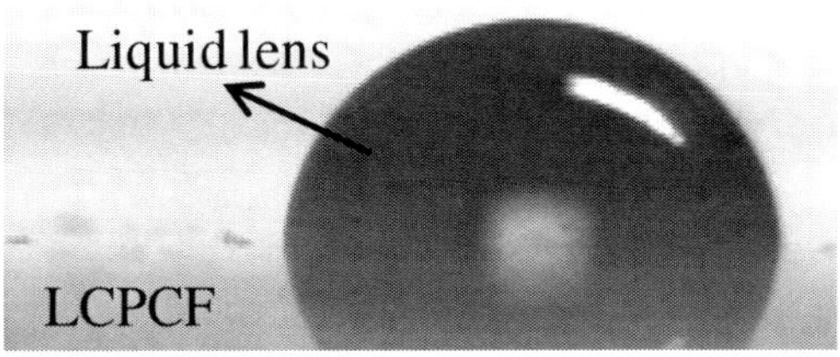

Figure 1-11. The recorded image of the liquid lens on LCPCF using CCD camera.

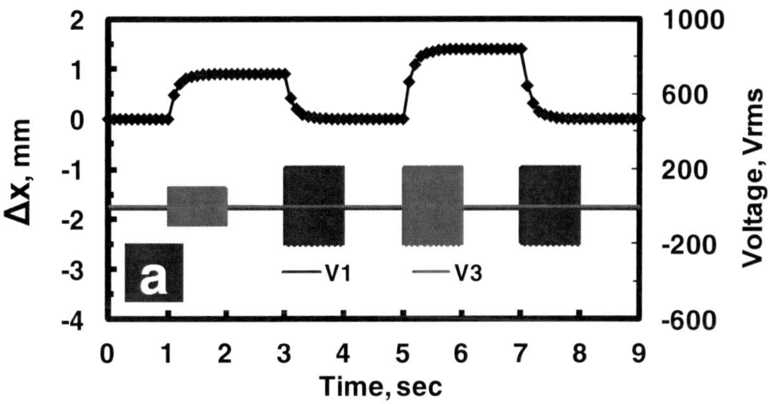

Figure 1-12. The moving distance (Δx) of liquid lens as a function of applied pulse voltages V_1 and V_3 [55].

The droplet can also be manipulated to move back and forth. Initially, the droplet was placed in the middle of the three electrode regions. When a square pulsed voltage was applied to V_3, the droplet moved toward electrode region 3 because the surface of the LCPCF at region 3 became more hydrophilic. Square-pulsed voltage was alternatingly applied to V_3 and V_1 with a time duration of 1 s and a frequency f of 1 kHz. The results are shown in Figure 1-

12. At t = 0 s, the droplet was placed in the middle of the LCPCF. At t = 1 s, the droplet moved a distance of 0.9 mm because of the applied pulsed voltage of $V_3 = 100\ V_{rms}$. When the voltage was turned off (i.e., $V_3 = 0$) at t = 2 s, the liquid lens stayed at the position of $\Delta x = 0.9$ mm. At t = 3 s, the droplet returned to the initial location (i.e., $\Delta x = 0$) because of the pulsed voltage of $V_1 = 200\ V_{rms}$ and $V_3 = 0\ V_{rms}$. At t = 4 s, the droplet stayed at the location when the voltages were turned off, indicating that the position of the droplet can be manipulated bistably by controlling the applied voltages.

1.3.2.4. Output Power and Efficiency

To measure the output power of the CPV system, the light source was set with a specific power density P_S impinged to the LC lens and then to the solar cell. The solar cell was connected to a power supply (Agilent, E3631A) and a multimeter (Agilent, 34401A). When voltages were applied to the LC lens, the current density and the voltage of the solar cell were measured. From the J-V curve, the P-V curve can be calculated because the output power density equals the multiplication of the current density and the voltage. From the P-V curve, the operating point of the solar cell at the voltage corresponding to the maximum output power density can be determined, and thus the operating voltage V_m and the operating current density J_m. The fill factor and the efficiency can also be calculated. The concentration ratio of the concentrator was modified by changing the applied voltage of the concentrator and the output power density and efficiency was measured using the same method, and the relation between the applied voltage and the output power density or efficiency can then be obtained.

1.4. CPV Systems Using Liquid Crystal Light Modulators

1.4.1. CPV Systems Using Liquid Crystal Lens

By adopting a liquid crystal (LC) lens into the concentrating photovoltaic (CPV) system, a CPV system with a steady electric output can be attained. The electrically tunable concentration ratio of the LC lens helps increase the photocurrent at low illuminations and prevent the effect of the series resistance at high illuminations. As a result, a CPV system with a high and static output power density can be realized by controlling the driving voltage of the LC lens. The concentration ratio of the LC lens can be switched from 1 to 9. The efficiency can be maintained at the maximum value under various illuminations [29].

1.4.1.1. Operating Principles

The operating principles of the proposed CPV system using a LC lens are illustrated in Figures 1-13(a) and 1-13(b). The system consists of an LC lens, a multijunction solar cell, and a DC-AC inverter. When sunlight passes through an LC lens, the LC lens modulates the number of photons per area (or photon-flux density). Two layers of LC were used to remove the polarization dependency. The photons are incident to the solar cell and are absorbed in the depletion region. The electron-hole pairs are excited in the depletion region when the energy of photons is larger than the diode band gap. The electrons are then excited from the valence band to the conduction band and generate the free electrons and holes, which is known as the photogeneration effect. The free electrons move toward the n-type region while the holes move toward the p-type region because of the built-in electric field of the depletion region, and the free electrons and holes form a photocurrent. The electric current (I) or DC resulting from the drifting electrons and holes can be obtained and converted to AC by a DC-AC inverter. Such an AC can provide electricity to a bulb and LC lens. The LC lens is a device in which the focusing properties change with the applied voltage because of the orientation of LC directors. Under strong illumination (Figure 1-13(a)), the distribution of the orientations of LC directors is controlled by the applied voltage and low photon-flux density then enters the solar cell to reduce the effects of the series resistance. Under weak illumination (Figure 1-13(b)), the distribution of the orientations of LC directors is adjusted to increase more photon-flux density entering the solar cell to enhance the output current (I). Assume that the mean photon-flux density of sunlight is φ_{onesun} and S is a sunlight factor that indicates the variation of sunlight depending on weather conditions. The photon-flux density of the incident light is $S \times \varphi_{onesun}$ before the sunlight passes through the LC lens. The number of photons that actually arrive to the accepted area of the solar cell can be modulated by the LC lens. The relation between the photon-flux density from the sunlight (i.e., $S \times \varphi_{onesun}$) and illumination accepted by the solar cell (φ_{input}) can be expressed as

$$\varphi_{input} = X(V_{LC}) \times S \times \varphi_{onesun}, \tag{1-16}$$

where X is the concentration ratio depending on the driving voltage of the LC lens (V_{LC}). Define a parameter $X_S(V_{LC})$ that is equal to $S \times X(V_{LC})$; $X_S(V_{LC})$ also indicates the total concentration ratio, which is the ratio of φ_{input} to φ_{onesun}. The illuminated solar cell is a current source with efficiency η, which is a ratio of the output electrical power to the power of the incident light. According to the photogeneration effect of the p-n junction, η can be expressed as:

$$\eta = \frac{J_{SC} \times V_{OC} \times FF}{h \times \nu \times \varphi_{input}}, \tag{1-17}$$

where J_{SC} is the short circuit current density or the photocurrent density, V_{OC} is the open circuit voltage, FF is the fill factor, h is the Planck constant (6.626×10^{-34} Joule-s), and ν is the mean frequency of the incident light. At a low photocurrent density, the effect of the series resistance is small enough to be neglected. When $X_S(V_{LC})$ increases, FF remains the same, but J_{SC} and V_{OC} increase. Consequently, η increases with the incident photon-flux density by a factor of $\left\{1 + \frac{k_B \cdot T}{q \cdot V_{OC,1}} \cdot \ln[X_S(V_{LC})]\right\}$, where k_B is the Boltzmann constant (1.38×10^{-23} Joule-K^{-1}), T is temperature in Kelvins, q is the electron charge, and $V_{OC,1}$ denotes the open circuit voltage at $\varphi_{input} = \varphi_{onesun}$. At the high photocurrent density, the effect of series resistance must be considered. Assume an equivalent series resistance is R_S and the power dissipation can be expressed as $J_{SC}^2 \times R_S$. Thus, the efficiency (η) depending on $X_S(V_{LC})$ can be expressed as

$$\eta[X_S(V_{LC})] = \eta_1 \times \left\{1 + \frac{k_B \times T}{q \times V_{OC,1}} \times \ln[X_S(V_{LC})]\right\} - \frac{[X_S(V_{LC})]^2 \times J_{SC,1}^{\ 2} \times R_S \times A}{X_S(V_{LC}) \times h \times \nu \times \varphi_{onesun}}, \tag{1-18}$$

where η_1 is the ideal efficiency under 1-sun illumination, A is the area of the solar cell, and $J_{SC,1}$ denotes the short circuit current density at $\varphi_{input} = \varphi_{onesun}$. When the concentration ratio of the LC lens or $X(V_{LC})$ is a fixed number and not tunable, $X_S(V_{LC})$ is a function of sunlight factor S. The variable η then changes with the sunlight factor. However, when the concentration ratio of the LC lens or $X(V_{LC})$ is electrically tunable, $X(V_{LC})$ can be adjusted by V_{LC} and $X_S(V_{LC})$ can then be maintained as a constant regardless of changes to S; therefore, η can be a constant at different values of S.

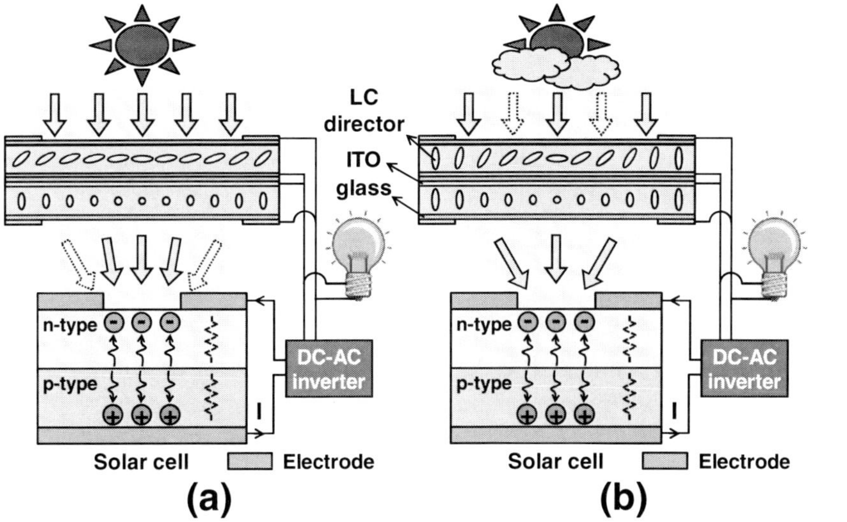

Figure 1-13. Operating principles of CPV system adopting a LC lens under (a) a strong illumination and (b) a weak illumination [29].

To obtain the highest η, we determine the maximum of $\eta[X_S(V_{LC})]$. Consequently, $\eta[X_S(V_{LC})]$ should satisfy Eq. (1-19):

$$\frac{d\eta[X_S(V_{LC})]}{dX_S(V_{LC})}=0\cdot \tag{1-19}$$

According to Eq. (1-19), the efficiency reaches a maximum when Eq. (1-20) is satisfied.

$$X_S(V_{LC})=\frac{h\times\nu\times\varphi_{onesun}\times\eta_1\times k_B\times T}{q\times V_{OC,1}\times {J_{SC,1}}^2\times R_S\times A}. \tag{1-20}$$

The result of the right side of Eq. (1-20) is a fixed number determined by the materials and the structure of the solar cell. Consequently, $X_S(V_{LC})$ is also an invariable constant. Therefore, the LC lens with an electrically switchable concentration ratio can not only help preserve the efficiency of the solar cell, but also achieve maximum efficiency.

1.4.1.2. Experimental Results and Discussion

To measure the concentration ratio of the LC lens, an unpolarized He-Ne laser (Melles Griot, 8144EU for $\lambda = 543.5$ nm and beam size of approximately 2 mm) was used. Instead of the solar simulator, here the unpolarized He-Ne laser was used to demonstrate the concept. The experimental trends of using the solar simulator and the laser are similar. The laser beam was impinged to the LC lens. We then measured the area ratio of the area of incident laser light to the area of the focused light at 25 cm behind the LC lens because the minimum focal length of the LC lens was 25 cm at $V_{LC} > 70\ V_{rms}$. This area ratio is also the concentration ratio X of the LC lens. The concentration ratio X of the LC lens as a function of the applied voltage V_{LC} is shown in Figure 1-14. In Figure 1-14, the concentration ratio increases from 1 to 4 with the applied voltage when the applied voltage is larger than the threshold voltage (approximately 20 V_{rms}). This is because the distribution of the orientations of LC directors changes with the applied voltage and results in the smaller focusing spot. The concentration ratio saturates at 4 after $V_{LC} > 70\ V_{rms}$. This is because the focal length did not change with the applied voltage as $V_{LC} > 70\ V_{rms}$. Therefore, the maximum concentration ratio of the LC lens is 4.

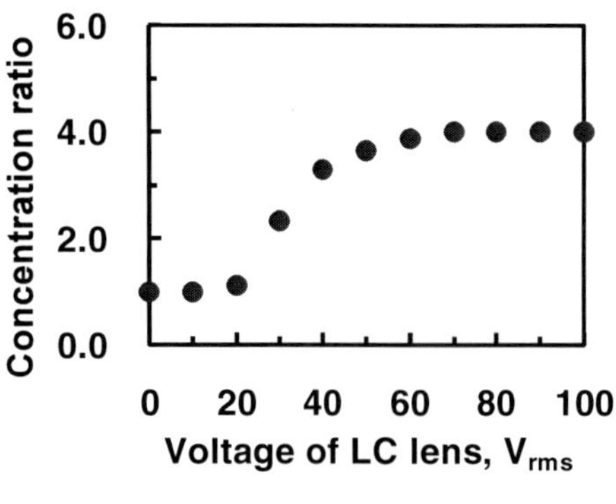

Figure 1-14. The concentration ratio (X) of the LC lens as a function of an applied voltage (V_{LC}).

To measure the efficiency of the CPV system in Figure 1-13(a), the same unpolarized He-Ne laser was used as a light source impinged to the LC lens and then to the solar cell. The structure of the solar cell was a GaInP/GaInAs/Ge triple junction (Arima, Model T3JG6F055011) with a wavelength range of 350 to 1800 nm and a diameter of 1 mm. The solar cell connected to a power supply (Agilent, E3631A) and a multimeter (Agilent,

34401A) was placed 25 cm behind the LC lens. A large area photodiode detector (New Focus, Model 2031) was used to measure the irradiance of light. When voltage was applied to the LC lens, the current and the voltage of the solar cell were measured. From the current-voltage curve, the power of the solar cell was calculated as a function of the voltage and the maximum power was determined (data not shown). The efficiency is the maximum power of the solar cell divided by the input power of the laser light. The efficiency was then plotted as a function of the applied voltage of the LC lens, as shown in Figure 1-15. The irradiance of the light source or sunlight factor (S) was also modified using attenuators to mimic variations in weather conditions. The variable S is defined as the ratio of the irradiance of the light source to the irradiance of 1 sun (approximately 1 mW/mm^2). In Figure 1-15, for a fixed S, at $V_{LC} < V_{th}$, the LC directors are not reoriented by the electric field, therefore the concentration ratio or incident number of photons does not change. As a result, the efficiency is unchanged. When $V_{th} < V_{LC} < 70\ V_{rms}$, the efficiency increases with V_{LC}, reaches a maximum, and then decreases. This is because the LC modulates the incident light and the concentration ratio or incident number of photons then increases with V_{LC}. Thus, the efficiency also increases. However, the power consumption induced by the effect of series resistance increases gradually as the incident number of photons increases. When the power consumption induced by the effects of series resistance is large enough to compete with the output power induced by the photocurrent of the solar cell, the efficiency reaches a maximum (approximately 4.5%) at $V_{LC} = V_{max}$ (<70 V_{rms}) and then begins to decrease. When $V_{LC} > 70\ V_{rms}$, the concentration ratio or incident number of photons remains the same, and the efficiency is unchanged.

In addition, the efficiency increases with S as $V_{LC} < V_{max}$ (Figure 1-15) because the number of photons increases as S increases. Therefore, the efficiency varies with S even though V_{LC} is fixed. This also indicates that the efficiency of the solar cell under a fixed concentration ratio of the LC lens (or a typical Fresnel lens case) is not constant, and the condition of the ambient light changes. The variable V_{max} depends on S. The variable V_{max} is 43 V_{rms} when S = 0.83, 47 V_{rms} when S = 0.80, 58 V_{rms} when S = 0.74, and 70 V_{rms} when S = 0.71. The maximum efficiencies are 4.5% similar, regardless of the sunlight condition, as shown in Figure 1-15. Therefore, according to the result of Figure 1-15, we can maintain the maximum efficiency of the CPV system by adjusting V_{LC} under various sunlight conditions.

From Figure 1-15, we plotted the maximum efficiency as a function of sunlight factor (black solid dots) as shown in Figure 1-16. The maximum efficiency is the efficiency at V_{max} in Figure 1-15 for a fixed S. Based on Eq. (1-18), we also calculated the maximum efficiency at various S values after considering the experimental parameters: $\eta_1 = 4.28\%$, T = 300 K, $q = 1.6 \times 10^{-19}$ C, $V_{OC,1} = 0.5$ V, $J_{SC,1} = 24$ A/m^2, $R_S = 40$ kΩ, A = 0.785 mm^2, $\nu = 5.52 \times 10^{14}$ Hz, $\varphi_{onesun} = 2.73 \times 10^{21}$ photons/s-m^2 , $X_S(V_{LC}) = 2.84$. The calculated results are shown in hollow dots in Figure 1-16. As the results indicate, the experimental results and calculated results of maximum efficiency agree.

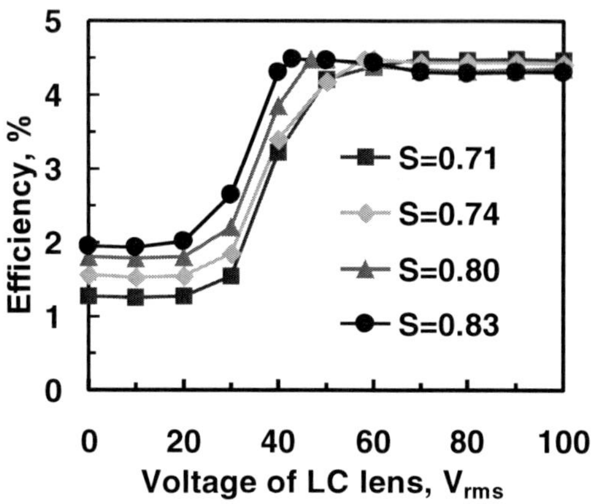

Figure 1-15. Efficiency as a function of an applied voltage of the LC lens at different S. S represents the ratio of the irradiance of the light source to the irradiance of 1 sun (~1mW/mm^2).

In addition, the maximum efficiency is nearly constant (approximately 4.5%), independent of S. This indicates that the CPV system can be operated at maximum efficiency by adjusting the applied voltage of the LC lens and that such efficiency does not change even though sunlight conditions change. In comparison, we also measured the efficiency of a CPV system using a glass lens whose concentration ratio was fixed (X was approximately 3.5) at different values of S, as shown in the red solid triangles in Figure 1-16. The efficiency of a CPV system using a glass lens changed with S, and was not constant. We then calculated the maximum efficiency at different values of S, as indicated by hollow red triangles in Figure 1-16, after considering $X_S = 0$ to 3.5 for S = 0 to 1. The experimental results agree with the calculated results. The efficiency of the CPV system using the glass lens changed with sunlight conditions; however, the efficiency of the CPV system using the LC lens is not only invariable, but also a maximum in the whole system. Therefore, the CPV system can be used at the highest efficiency under all types of sunlight conditions by using LC lenses.

The power consumption of the LC lens is lower than that of the entire CPV system. The measured current of the LC lens was 0.19 μA and the maximum voltage of the LC lens in the CPV system was 70 V_{rms}. Therefore, the power consumption of the LC lens is approximately 6.65 μW. The output power of the CPV system was also calculated. The output power of the CPV system adopting the LC lens ranged from 0 to 141 μW when S changed from 0 to 1 for the area of the LC lens of 3.14 mm^2 and the efficiency of 4.5%; conversely, the output power of the CPV system using the glass lens ranged from 0 to 123 μW (S = 0 to 1) within the area of the glass lens of 2.75 mm^2, and the variable efficiency ranged from 0% to 4.5%. The output power of the CPV system using the LC lens was larger than that of the CPV system using the glass lens. To further improve the output power of the CPV system using the LC lens, the power consumption of the LC lens can be diminished by reducing the voltage used and designing a new LC lens structure; the concentration ratio can be increased by enlarging the diameter of the LC lens.

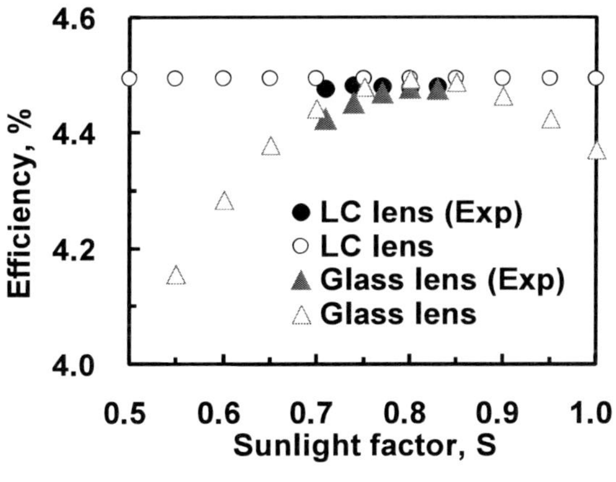

Figure 1-16. The maximum efficiency as a function of the sunlight factor. Black solid dots stand for the experimental result and hollow black dots stand for calculated results in the CPV system adopting the LC lens. Red solid triangles stand for the experimental result and hollow red triangles stand for calculated results in the CPV system adopting the glass lens.

1.4.1.3. Wavelength Effect

We have discussed the power of a CPV system under monochromatic illumination. However, because sunlight is a broadband light source which has a spectrum of approximately 300–1800 nm, we must also discuss the CPV system using the LC lens under chromatic illumination. In Section 1.2.2, we discussed the relationship between the refractive indices and the wavelength of incident light. For example, the dispersion of the MLC-2070 is shown in Figure 1-17. Furthermore, how the focal length of the LC lens depends on the distribution of the refractive index was derived in Section 1.3.2. Thus, the focal length and the concentration ratio of the LC lens depending on the wavelength can be derived, as shown in Figures 1-18 and 1-19. Figure 1-19 shows that the LC lens has a maximum concentration ratio at various applied voltages at different wavelengths. The total output power under chromatic illumination is a summation of the output power under the monochromatic illuminations of different wavelengths, which are also dependent on the properties of the solar cell material. Thus, for different types of solar cell, a suitable voltage can be applied to the LC lens to obtain a maximum output power with a specific concentration ratio.

1.4.1.4. Distance Effect

The concentration ratio can also be calculated from the focal length by geometrical optics. The experimental setup is shown in Figure 1-20.

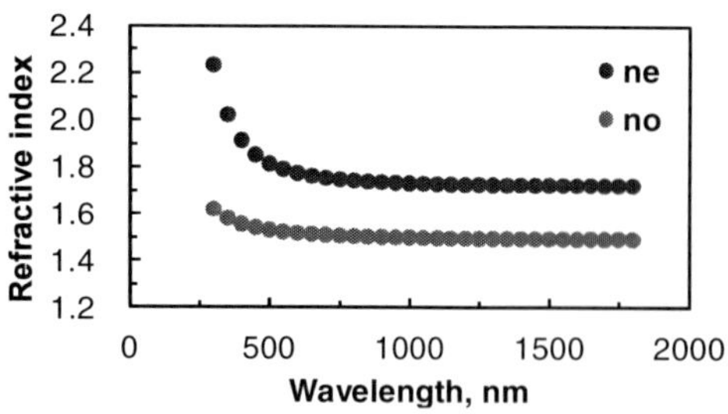

Figure 1-17. Birefringence depends on wavelength.

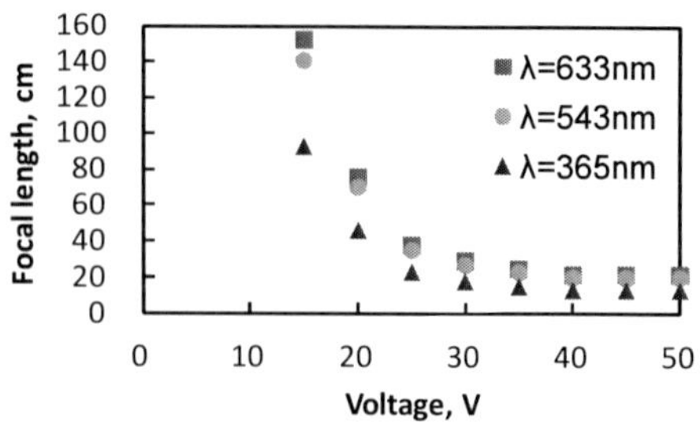

Figure 1-18. Focal length depends on wavelength.

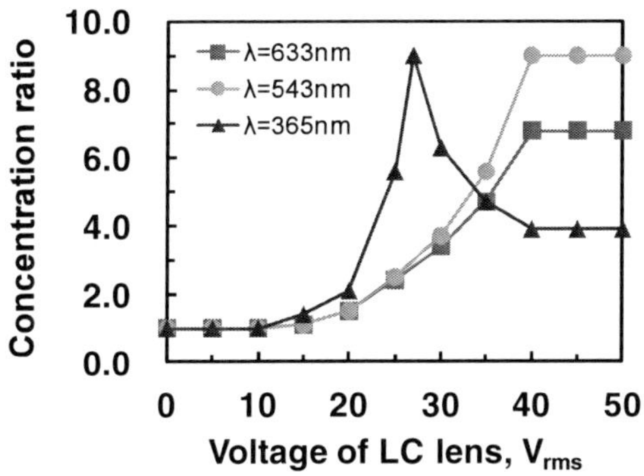

Figure 1-19. Concentration ratio depends on wavelength at d=20 cm.

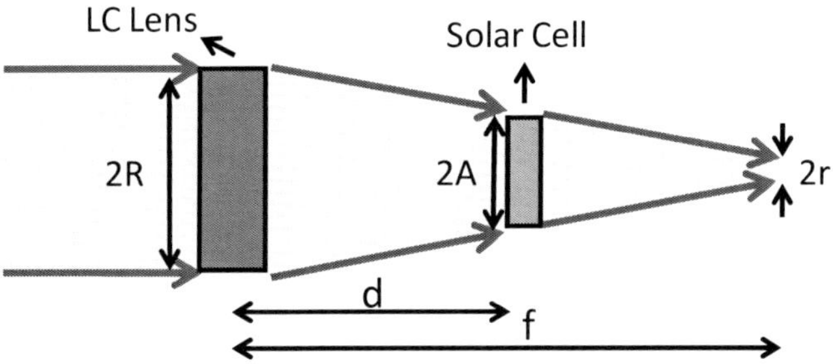

Figure 1-20. Calculation of concentration ratio.

In Figure 1-20, assume the radius of LC lens is R, the radius of the focal point at the focal length of the LC lens is r, the radius of the illumination area at the plane of solar cell is A, the distance between the LC lens and the solar cell is d, then by geometrical optics A as a function of d can be expressed as:

$$A = R - \frac{d \cdot (R - r)}{f} \tag{1-21}$$

and the concentration ratio X can be expressed as:

$$X = \frac{R^2}{A^2} \tag{1-22}$$

Eq. (1-21) and (1-22) shows that when the radius of LC lens R and the radius of the focal point at the focal length of the LC lens r are fixed, the concentration ratio X is dependent on the system distance d. When the system distance d increases, the applied voltage that the maximum concentration ratio appears decreases. We also know that the concentration ratio is dependent on the wavelength in the previous experimental results. Therefore, the effects of system distance d and the wavelength can be summarized: When d is small, the system is more compact, but only light of short wavelength can reach the maximum concentration ratio of the LC lens, as shown in Figure 1-21; When d is large, the system is bulky, but a large range of incident light of different wavelength can reach the maximum concentration ratio with different applied voltage of LC lens, as shown in Figure 1-22.

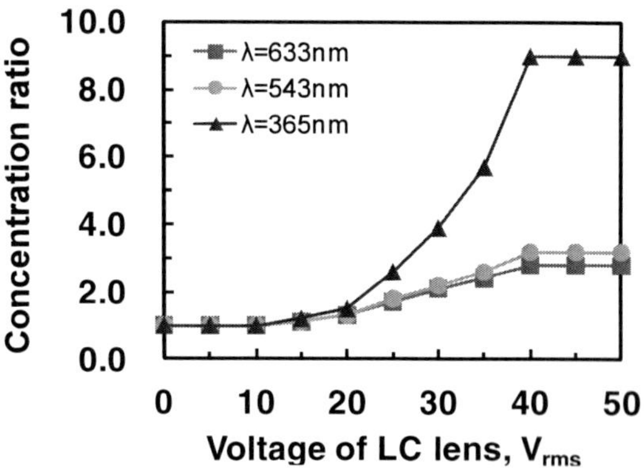

Figure 1-21. Concentration ratio depends on wavelength at d=13.2 cm.

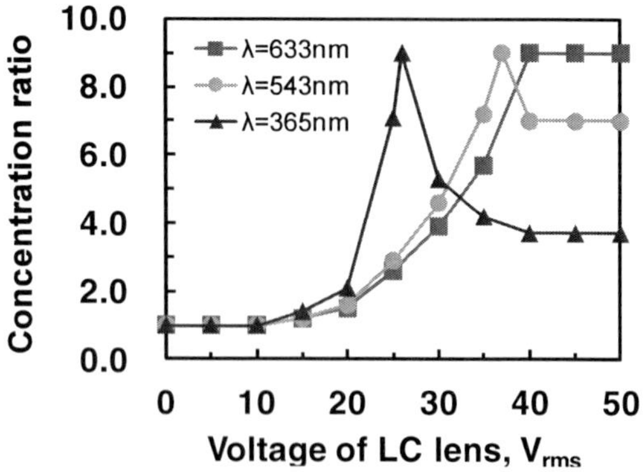

Figure 1-22. Concentration ratio depends on wavelength at d=21.6 cm.

1.4.1.5. System Limitations

Eq. (1-20) has shown the condition that the system has its maximum efficiency. Assume S=1 and X_i is a solution of Eq. (1-20), that means under one-sun illumination, the system has it maximum efficiency when the LC lens has a concentration ratio X_i. If the range of the concentration ratio of the LC lens is from X_1 to X_2 ($X_1 < X_i < X_2$), then the range of the intensity of the incident light that the efficiency can be maintain at the maximum by LC lens can be derived as:

$$\frac{X_{ideal}}{X_2} < S < \frac{X_{ideal}}{X_1} \qquad \text{1-23)}$$

Therefore, if we want to decrease the lower bound in Eq. (1-23) ($\frac{X_{ideal}}{X_2} \downarrow$), we have to increase X_2, that means the diameter of the LC lens should be increased and the diameter of the focal spot at the focal plane should be decreased. On the other hand, if we want to increase the upper bound in Eq. (1-23) ($\frac{X_{ideal}}{X_1} \uparrow$), we have to decrease X_1, that means the maximum focal length should be as long as possible. In this case, the performance of the system is limited by the aperture size of the LC lens, which results in an acceptable range $0.7 < S < 0.83$ of S, as shown in Figure 1-16. We have also used a liquid lens, which has a diameter of 10 mm, to do the same experiment. We found that the system work well, and the acceptable range of S is around 9 times comparing to the LC lens. This is because the concentrator with a larger aperture size has a larger maximum concentration ratio. Thus, to improve the performance of LC lens, we have to increase the aperture size of LC lens and preserve a small focal spot at the same time.

We have also considered the effect of the series resistance. According to Eq. (1-20), when the series resistance R changes, the concentration ratio where the maximum efficiency appears will also changes. The experimental result is shown in Figure 1-23.

In Figure 1-23, we found that the concentration ratio where the maximum efficiency appears increases as the series resistance decreases. The series resistance is dependent on the material and the fabrication process of the solar cell. A better solar cell (a more efficient solar cell) has a less series resistance. Thus, when the performance of the solar cell itself is enhanced, the maximum concentration ratio of the LC lens should also be increased to ensure the system can reach its maximum efficiency.

In conclusion, a CPV system with a steady electrical output adopting an electrically tunable concentration ratio of the LC lens was demonstrated. By controlling the applied voltage of the LC lens in order to adjust the number of incident photons in an area, the CPV system can be operated in the fixed and maximum efficiency under different ambient illuminations. The concept is not only limited to LC lens, but also can apply to other optical elements whose concentration ratio is tunable, such as liquid lenses, spatial light modulators, and light-induced mechanical stress on a polymer film.

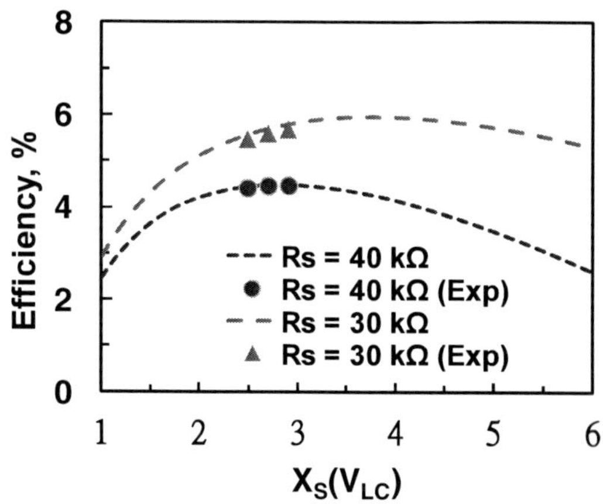

Figure 1-23. Efficiency depends on concentration ratio with different series resistance.

1.4.2. CPV Systems Using Liquid Crystal and Polymer Composite Film

In this section, we introduce a CPV system using droplet manipulation on an LC and polymer composite film (LCPCF). This CPV system adopted a droplet as a liquid lens on the LCPCF whose curvature is not only tunable, but whose position is also bistably switchable based on LC orientations on the LCPCF. The change of curvature of the liquid lens results in a tunable concentration ratio that helps to increase photocurrent at low illuminations and prevents series resistance at high illuminations. Moreover, because the position of the liquid lens is changeable, it can be modified as the sun moves.

Therefore, the output power of such a system is steady regardless of the sunlight condition and the angle of incident light [55].

1.4.2.1. Operating Principles

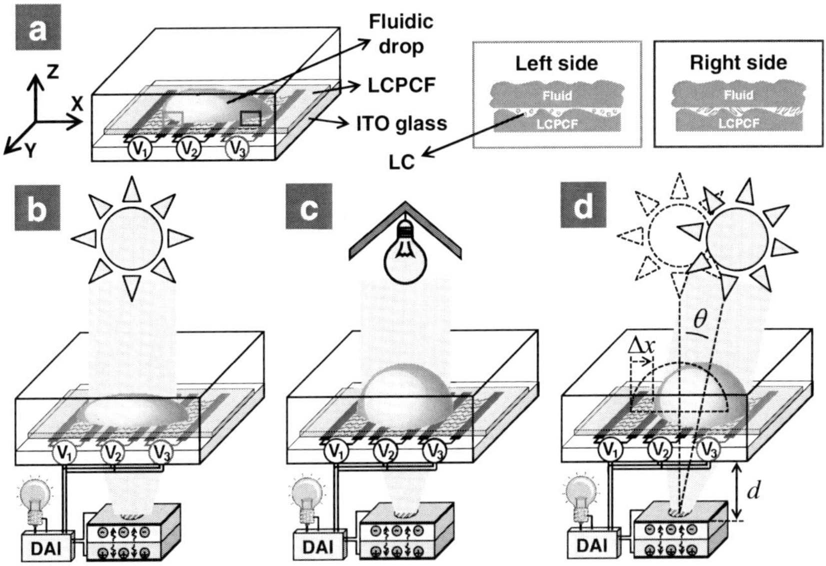

Figure 1-24. Operating principles of the droplet manipulation on LCPCF for CPV system [55].

The operating principles of the proposed CPV system are illustrated in Figures 1-24(a), 1-24(b), 1-24(c), and 1-24(d). In Figure 1-24(a), the mechanism of the liquid lens is illustrated based on the droplet manipulation on the LCPCF. A fluidic drop was placed on the LCPCF and a patterned ITO glass substrate with three electrode regions was set underneath, as shown in Figure 1-24(a). The patterned ITO glass substrate provided fringe electric fields at the surface of the LCPCF. The ITO electrodes on the glass substrate were etched with interdigitated chevron patterns, shown as the zigzag electrodes in Figure 1-24(a). The zigzag ITO strips have corner angles of 150°. The width and gap of the electrode strips were 4 μm and 14 μm, respectively. To manipulate a fluidic drop on the LCPCF, the three regions of the interdigitated chevron electrodes were patterned identically, as shown in Figure 1-24(a). The fluidic drop was originally located in the middle of the three electrode regions. When pulsed voltage was applied to the right side of the LCPCF (i.e., V_3), as in Figure 1-24(a), for example, the fluidic drop moved to the right because the right region is more hydrophilic or the surface free energy of the right region is higher because of the tilts of the LC orientation, as shown in the two square insets of Figure 1-24(a). The relationship between the electrically tunable surface free energy of the LCPCF and the LC orientation was discussed previously. When identical pulsed voltages were applied to V_1, V_2, and V_3, the curvature of the fluidic drop changed without moving. When pulsed voltages were alternatingly applied to V_3 and V_1, the fluidic drop moved to the right region and the left region accordingly. As a result, the fluidic drop on the LCPCF can function as an electrically tunable focusing liquid lens because of the change of the curvature, and control the shift of focusing spot along the focusing plane, which is typically the plane of the solar cell located based on the change of the position of the fluidic drop. This also indicated that the liquid lens on the LCPCF can function not only as a concentrator but also as a sun tracker. Then the liquid lens in Figure 1-24(a) is placed on the top of a multijunction solar cell as shown in Figures 1-24(b), 1-24(c), and 1-24(d). In Figures 1-24(b) and 1-24(c), when the sunlight passes through the liquid lens, the curvature of the liquid lens can be manipulated by applying various homogeneous voltages (i.e., $V_1 = V_2 = V_3 \equiv V_{total}$) to modulate the number of photons over an area and over a period of time (i.e., photon-flux density) to the solar cell. The absorbed photons in the depletion region of the solar cell generate free electrons and holes to form a photocurrent that can be further converted to AC by using a DC-AC inverter. Such an alternating current provides electricity to the LCPCF and a bulb (or other device). The radius of the curvature of the liquid lens is large, which corresponds to a lower concentration ratio as a concentrator. A low concentration ratio of the liquid lens was used under strong sunlight to prevent series resistance at high illuminations. Under weak illumination, such as an indoor environment as depicted in Figure 1-24(c), the concentration ratio of the liquid lens was increased by decreasing V_{total} to increase the photocurrent at low illumination. Thus, the conditions in Figures 1-24(b) and 1-24(c) are able to have the same electric output power. However, based on the movement of the sun, the oblique sunlight results in a decrease of the output power of the solar cell. To compensate for the oblique effects of sunlight, we shifted the position of the liquid lens to shift the oblique sunlight back to the solar cell according to the droplet manipulation on the LCPCF, as shown in Figure 1-24(d). Although oblique sunlight changes the irradiance because of the change of the ambient conditions (i.e., indoor or outdoor), the position of the liquid lens and the radius of curvature can be modified simultaneously to maintain the generated output power.

To prove the operating principles, we assume the oblique angle of the sun is θ with respect to the normal direction of the solar cell, and is θ' with respect to the normal direction of the liquid lens, as shown in Figures 1-25(a) and 1-25(b). The distance between the liquid lens and the solar cell is d. Consequently, the focused sunlight shifts a distance of $d \times \tan\theta'$, as shown in Figure 1-25(a). To compensate for the oblique effects of sunlight, the liquid lens moves a distance of $\Delta x(V_3)$ by applying a pulsed voltage of V_3, as depicted in Figure 1-25(b). From Figure 1-25(b), $\Delta x(V_3)$ equals $d \times \tan\theta$. Assume the mean photon-flux density (φ_{onesun}) is approximately 1 mW/mm^2 when the sun is above the liquid lens and the solar cell, and S is the sunlight factor that indicates the variation of sunlight depending on weather conditions.

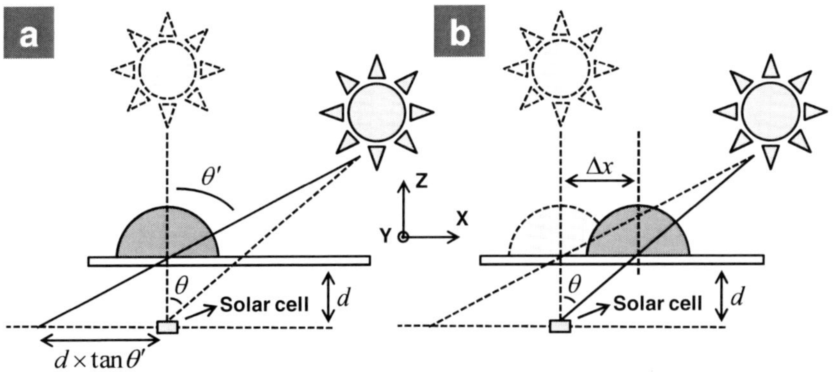

Figure 1-25. The illustration of the sun-tracking mechanism by using the movement of liquid lens. (a) When the incident angle of sunlight change, the focusing spot shift a distance of $d \times \tan\theta'$. (b) To shift the focusing spot back to the solar cell, the liquid lens shifts a distance of Δx [55].

The photon-flux density (φ_{input}) accepted by the solar cell at the oblique incidence is expressed as

$$\varphi_{input} = \varphi_{onesun} \times S \times \cos\theta \times C(V_{total}), \quad (1\text{-}24)$$

where C is a voltage-dependent concentration ratio of the liquid lens, defined as the area ratio of the liquid lens to the solar cell at the normal incidence. The oblique effect of sunlight is considered by multiplying cosθ in Eq. (1-24). In our previous study, to achieve the maximum of the output electrical power density, φ_{input} should be satisfied in Eq. (1-25).

$$k_B \times T/q \times FF \times \left\{ \ln\left[\frac{\varphi_{input}}{\varphi_{onesun}}\right] + 1 \right\} - 2 \times \frac{\varphi_{input}}{\varphi_{onesun}} \times J_{SC,1} \times R_S + FF \times V_{OC,1} = 0, \quad (1\text{-}25)$$

where k_B is the Boltzmann constant (1.38×10^{-23} Joule-K^{-1}), T is temperature in Kelvins, q is the electron charge, FF is the fill factor, R_S is the series resistance, and $J_{SC,1}$ and $V_{OC,1}$ denote the short circuit current and open circuit voltage at $\varphi_{input} = \varphi_{onesun}$, respectively. The solution of φ_{input} in Eq. (1-25) is a constant, defined as φ_S, because the other parameters in Eq. (1-25) are fixed. From Eq. (1-24), φ_S equals $\varphi_{onesun} \times S \times \cos\theta \times C(V_{total})$. At a normal incidence (i.e., $\theta = 0$), we adjusted the concentration ratio $C(V_{total})$, satisfying $C(V_{total}) = \varphi_S / (\varphi_{onesun} \times S)$ to maintain the maximum of P_{out} regardless of the sunlight condition (the value of S). Under an oblique incidence (i.e., $\theta \neq 0$), the liquid lens was moved a distance of $\Delta x(V_3)$ ($= d \times \tan\theta$) by applying voltage V_3 and the concentration ratio $C(V_{total})$ was then readjusted to satisfy $C(V_{total}) = \varphi_S / (\varphi_{onesun} \times S \times \cos\theta)$. Thus, P_{out} remains a maximum. Therefore, manipulating the concentration ratio and the location of a droplet on the LCPCF can be simultaneously used as a concentrator and a sun tracker for CPV systems for obtaining a steady and maximum output electric power.

1.4.2.2. Experimental Results and Discussion

To demonstrate the concept of the proposed CPV system, we prepared the LCPCF with the liquid lens. To fabricate the LCPCF, an empty cell consisting of a patterned ITO glass substrate and a glass substrate coated with a mechanically buffered polyimide layer was prepared. The cell was then filled with a mixture containing a nematic LC mixture E7 (Merck) and a liquid crystalline monomer (4-(3-acryloyloxypropyloxy)-benzoic acid 2-methyl-1, 4-phenylene ester) at 70:30 wt% ratios, and the cell was exposed to UV light. After phase separation and photopolymerization, the top glass substrate was peeled off by using a thermal-releasing process. A 6-μL drop of DI water was dropped on the solidified LCPCF to form a liquid lens, after which the device was assembled using another glass substrate to avoid the evaporation of the liquid lens.

To measure the output power density of the CPV system, a white light source (Taiwan Fiber Optics, LSH-150F, with an irradiance of approximately 1 mW/mm^2) was used to mimic sunlight and illuminate the liquid lens. The structure of the solar cell was a GaInP/GaInAs/Ge triple junction (Arima, Model T3JG6F055011) with a wavelength range of 350–1800 nm and a diameter of 1 mm. The solar cell was placed at d = 3.2 mm behind the liquid lens and connected to a power supply (Agilent, E3631A) and a multimeter (Agilent, 34401A), as depicted in Figure 1-24(b). The current-voltage (I-V) curve of the solar cell was measured and the output power density was then calculated as a function of the voltage. The maximum output power density defined as P_{out} was chosen based on the relationship between the output power density and the voltage. We measured P_{out} as a function of the concentration ratio of the liquid lens by applying voltage under various sunlight conditions (S), as shown in Figure 1-26(a). The value of S is defined as the ratio of the irradiance of the light source to the irradiance of 1 sun (approximately 1 mW/mm^2). In Figure 1-26(a), the P_{out} increases with the concentration ratio first because of the photovoltaic effect and then decreases after reaching a maximum because of series resistance. The concentration ratios necessary to achieve the maximum P_{out} (approximately 325 μW/mm^2) for S = 1, 0.85, and 0.7 are 6.4, 7.6, and 9.0, respectively. This indicates that the solar cell generates the highest output power density (i.e., 325 μW/mm^2 in Figure 1-26(a)) under various sunlight conditions, if the concentration ratio of the liquid lens is adjusted without moving the location of the liquid lens (i.e., Δx = 0 in Figures 1-24(d) and 1-25(b)).

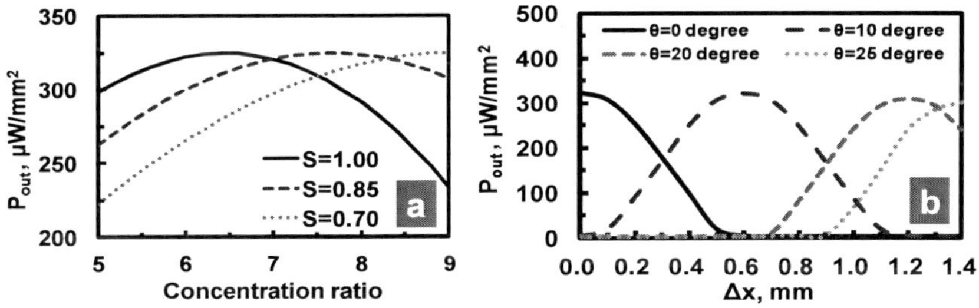

Figure 1-26. (a) Output power density as a function of the concentration ratio under different sunlight conditions. (b) Output power density as a function of the moving distance at different incident angles when S=1 and C=6.4 [55].

The location of the liquid lens was subsequently moved to observe the changes of P_{out} under different angles of sunlight (i.e., θ in Figures 1-24(d) and 1-25(b)). The variable P_{out} as a function of moving distance is shown in Figure 1-26(b). At normal incidence (i.e., $\theta=0°$), the output power density decreased with the moving distance Δx because the focusing spot of sunlight after passing through the liquid lens shifted away from the solar cell. At $\theta=10°$, P_{out} had a maximum of 322 $\mu W/mm^2$ when $\Delta x = 0.6$ mm. At $\theta=20°$, P_{out} had a maximum of 310 $\mu W/mm^2$ when $\Delta x = 1.2$ mm. At $\theta=25°$, P_{out} had a maximum of 302 $\mu W/mm^2$ when $\Delta x = 1.4$ mm. Consequently, the position of the liquid lens can be changed to trace the sun to obtain the maximum output power density. From Figure 1-26(b), we plotted the moving distance Δx when P_{out} is maximal as a function of an incident angle θ, as shown by black dots in Figure 1-27. We also plotted theoretical Δx as a function of $d \times \tan\theta$ (d was 3.2 mm in our experiments), as shown by the black dotted line in Figure 1-27. The experimental results and theoretical results were similar. Therefore, regardless of the location of the sun, the local maximum P_{out}, as shown in Figure 1-26(b), can be obtained by moving the liquid lens. We also plotted the local maximum P_{out} in Figure 1-26(b) as a function of an incident angle, as shown by the green dotted line with squares in Figure 1-27. The local maximum P_{out} decreased to approximately 93% when the incident angle changed from 0 to 25°. This is because the oblique light resulted in the decrease of the photon flux density (i.e., photon numbers per unit area per unit time). For comparison, we also measured P_{out} as a function of the incident angle without the location change of liquid lens (the red dotted line with triangles). The results revealed that the output power density dropped dramatically at $\theta > 5°$ at $\Delta x = 0$. However, the output power density slowly decreased when the location of the liquid lens was adjusted. Therefore, the output power density decreased by only 93% when the incident angle was 25°.

To obtain a stable output power density without a 93% decrease as the incident angle changes, the liquid lens can be moved to trace the sun first and the concentration ratio of the liquid lens can then be adjusted to improve the output power density. In the experiments, the angle of the incident light (i.e., θ) was controlled and the liquid lens Δx was then moved according to the black dotted line in Figure 1-27.

Subsequently, the output power density was measured when changing the concentration ratio of the liquid lens. The output power density as a function of the concentration ratio of the liquid lens is depicted in Figure 1-28. The figure indicates that the output power density can attain local maximum by adjusting the concentration ratio of the liquid lens at various incident angles, which requires moving the liquid lens to different locations.

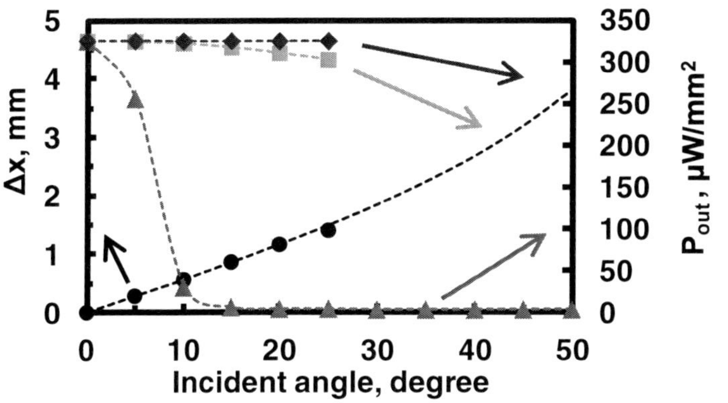

Figure 1-27. The moving distance as a function of incident angle (black dots). The output power density as a function of incident angle when Δx = 0 mm (red triangles). Output power density as a function of incident angle with the manipulation of the moving distance Δx of the liquid lens (green squares). The output power density as a function of incident angle with the manipulation of both of moving distance Δx and the concentration ratio C (blue diamonds) [55].

Moreover, the local maximums of four curves were identical at approximately 325 μW/mm^2. We also plotted P_{out} as a function of the incident angle after manipulating both the position and the concentration ratio of the liquid lens, as indicated by the blue dotted line with diamonds in Figure 1-27. Therefore, the position of the liquid lens can be manipulated to trace the sun and then change the concentration ratio of the liquid lens to obtain stable output power density. In other words, manipulating the motion and the curvature of the liquid lens enabled the attainment of a stable output power of the solar cell despite the sun's movement.

According to the experimental results in Figure 1-28, the maximum φ_{input} in the system was approximately 6.4 mW/mm^2, which was calculated by inputting the experimental parameters into Eq. (4-9): S = 1, θ = 0°, and C = 6.4. We can also estimate the theoretical value of the maximum φ_{input} by inputting the related parameters into Eq. (4-10): T = 300 K, q = 1.6 × 10^{-19} C, FF = 0.85, $J_{SC,1}$ = 4 mA, $V_{OC,1}$ = 2.37 V, and R_S = 40 kΩ. The theoretical φ_{input} (i.e., φ_S) was approximately 6.5 mW/mm^2, which is similar to the experimental result (6.4 mW/mm^2). This also indicates that the liquid lens using droplet manipulation on the LCPCF is not only a concentrator but also a sun tracker for stabilizing and maximizing the output power density in the CPV system. To operate the liquid lens on the LCPCF, we had to apply voltage that also consumed energy. Regarding the power consumption of the liquid lens on the LCPCF, the measured power consumption of the LCPCF was less than 1 $\mu W/mm^2$ at any applied voltage of the LCPCF. Compared with the output power density of approximately 325 $\mu W/mm^2$, the energy gain is larger than approximately 300 $\mu W/mm^2$ and the power consumption of the liquid lens on the LCPCF is negligible.

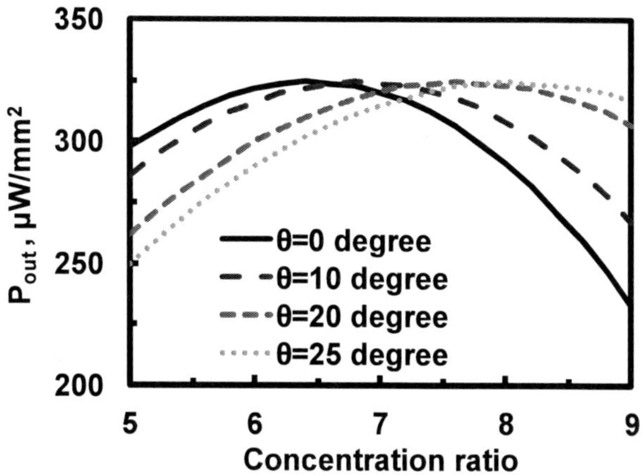

Figure 1-28. The output power density as a function of concentration ratio at different incident angles and at related moving distance Δx. The relation between incident angles and the moving distance is plotted in the black dots in Figure 1-27 [55].

CONCLUSION AND FUTURE SCOPE

In this chapter, we developed electrically tunable liquid crystal concentrators for concentrating photovoltaic systems. We also extended the operating principles of the concentrators to be used in other applications, such as in image stabilization systems. On the basis of controlling the concentration ratio and tracking the direction of the concentrators, we can realize a CPV system with a static and maximum output power density under different ambient illumination conditions, because the electrically tunable concentrator helps increase the photocurrent at low illuminations and prevent series resistance at high illuminations, while simultaneously tracking the sun.

The main objective of this chapter was to highlight the feasibility of adopting tunable liquid crystal devices in photovoltaic systems. Conventional concentrating photovoltaic systems use concentrators with a fixed concentration ratio, which cannot satisfy the solar cell when illumination conditions change. A tunable concentrator enables the solar cell to function at a high efficiency under various illumination conditions and increases the total output power during a period of time. The small increase in output power in our experimental results was because of the limitation of the performance of the solar cell. Thus, the performance of this mechanism can improve, depending on the development of the solar cell itself. The liquid crystal concentrator and sun tracker require only a negligible amount of output power to function, and can be applied to all types of solar cell. The mechanisms of the liquid crystal concentrator and sun tracker can also be applied to other methods for improving efficiency, such as antireflection coatings, photon recycling, or light trappings. We can also further improve performance by integrating more mechanisms into a single device, such as a film that can simultaneously concentrate light, track the sun, and minimize reflections.

The concept presented in this chapter is not limited to liquid crystal devices, but can also be applied to other tunable optical elements, such as liquid lenses, spatial light modulators, and beam steering devices.

REFERENCES

[1] D. Anderson, *Clean Electricity from Photovoltaics*, eds. M.D. Archer and R.D. Hill (London: Imperial College Press, 2011).

[2] M. A. Green, *Conf. Record 21st IEEE Photovoltaic Specialists Conf.*, 1-7 (1990).

[3] J. N. Shive, *Semiconductor Devices* (Van Nostrand, 1959).

[4] C. A. Vincent, *Modern Batteries* (Arnold, 1997).

[5] M. Wolf, *Proc. 25th Power Sources Symposium, 1972*. In Solar cells, ed. c.E. Backus (IEEE Press, 1976).

[6] J. Nelson, *The Physics of Solar Cells* (Imperial College Press, London, 2003).

[7] G. L. Araujo and A. Marti, *Solar Energy Materials and Solar Cells* 33, 213 (1994).

[8] A. de Vos, *Endoreversible Thermodynamics of Solar Energy Conversion* (Oxford University Press, 1992).

[9] R. Gottschalg, *The Solar Resource and the Fundamentals of Radiation*

for Renewable Energy Systems (Sci-Notes, Oxford, 2001).
[10] C. H. Henry, *J. Appl. Phys.* 51, 4494 (1980).
[11] W. Shockley and H. J. Queisser, *J. Appl. Phys.* 32, 510 (1961).
[12] P. Wuerfel, J. *Phys.* C15, 3697 (1982).
[13] A. Goetzberger and W. Greubel, *Appl. Phys. Letts.* 14, 123 (1977).
[14] M. A. Green, *Silicon Solar Cells: Advanced Principles and practice* (Sydney: Centre for Photovoltaic Engineering, 1995).
[15] J. E. Parrott, Sol. En. Materials Sol. Cells 30, 221 (1993).
[16] R. A. Sinton, *Solar Cells and their Applications*, ed. L.D. Partain (Wiley, 1995).
[17] R. M. Swanson and R. A. Sinton, *Advances in Solar Energy* 6, 427 (1990).
[18] L. Stalmans J. Poortmans, H. Bender, M. Caymax, K. Said, E. Vazsonyi, J. Nijs,and R. Mertens, *Progr. Photovoltaics* 6, 233 (1998).
[19] W. G. J. H. M. van Sark, A. Meijerink and R. E. I. Schropp, *Third Generation Photovoltaics* (Vasilis Fthenakis, 2012)
[20] C. W. Oseen, *Trans. Faraday Soc.* 29, 883 (1933).
[21] H. Zocher, *Trans. Faraday Soc.* 29, 945 (1933).
[22] F. C. Frank, *Discuss Faraday Soc.* 25, 19 (1958).
[23] P. J. Collings and M. Hird, *Introduction to Liquid Crystals Chemistry and Physics* (Taylor & Francis, 1997).
[24] S. Chandrasekhar, *Liquid Crystals 2nd Edition* (Cambridge Univ. Press, 1992).
[25] G. Barbero and L. R. Evangelista, *An Elementary Course on The Continuum Theory for Nematic Liquid Crystals* (World Scientific, 2001).
[26] A. Yariv and P. Yeh, *Optical Waves in Crystals* (Wiley, 2003).
[27] B. E. A. Saleh and M. C. Teich, *Fundamental of Photonics*, (Wiley, 1991).
[28] N. Fraval and F. Berier, *Proc. SPIE* 7930, 793009-1 (2011).
[29] Y. S. Tsou, Y. H. Lin, and A. C. Wei, *IEEE Photon. Technol. Lett.* 24, 2239 (2012).
[30] Y. H. Lin and H. S. Chen, *Opt. Express* 21, 9428 (2013).
[31] H. C. Lin, N. Collings, M. S. Chen, and Y. H. Lin, *Opt. Express* 20, 27222 (2012).
[32] Y. H. Lin and M. S. Chen, *J. Display Technol.* 8, 401 (2012).
[33] H. C. Lin and Y. H. Lin, *Opt. Express* 20, 2045 (2012).
[34] H. C. Lin, M. S. Chen, and Y. H. Lin, *J. Nonlinear Opt. Phys. Mater.* 20, 477 (2011).
[35] H. C. Lin, M. S. Chen, and Y. H. Lin, *Trans. Electr. Electron. Mater.* 12,

234 (2011).
[36] Y. H. Lin, M. S. Chen, and H. C. Lin, *Opt. Express* 19, 4714 (2011).
[37] H. C. Lin and Y. H. Lin, *Appl. Phys. Lett.* 98, 083503 (2011).
[38] H. C. Lin, M. S. Chen and Y. H. Lin, *Mol. Cryst. Liq. Cryst.* 544, 150 (2011).
[39] H. C. Lin and Y. H. Lin, *Appl. Phys. Lett.* 97, 063505 (2010).
[40] H. C. Lin and Y. H. Lin, *J. Jpn. Appl. Phys.* 49, 102502 (2010).
[41] D. K. Yang and S. T. Wu, *Fundamentals of Liquid Crystal Devices* (Wiley, West Sussex, 2006)
[42] S. Sato, *Jpn. J. Appl. Phys.* 18, 1679 (1979).
[43] Y. H. Lin and Y. S. Tsou, *J. Appl. Phys.* 110, 114516 (2011).
[44] M. Ye and S. Sato, *Jpn. J. Appl. Phys.* 41, L571 (2002).
[45] M. Ye and S. Sato, *Jpn. J. Appl. Phys.* 42, 6439 (2003).
[46] M. Ye, B. Wang, M. Yamaguchi, and S. Sato, pn. *J. Appl. Phys.* 47, 4597 (2008).
[47] M. Ye, B. Wang, and S. Sato, *Opt. Express* 16 ,4302 (2008).
[48] Y. H. Lin, H. Ren, Y. H. Wu, S. T. Wu, Y. Zhao, J. Fang, and H. C. Lin, *Opt. Express* 16, 17591 (2008).
[49] Y. H. Lin, H. Ren, Y, H, Wu, Y. Zhao, J. Fang, Z. Ge and S. T. Wu, *Opt. Express* 13, 8746 (2005).
[50] Y. H. Lin, H. Ren, S. Gauza, Y. H. Wu, Y. Zhao, J. Fang, and S. T. Wu, J. *Display Technol.* 2, 21 (2006).
[51] Y. H. Lin, J. K. Li, T. Y. Chu, and H. K. Hsu, *Opt. Express* 18, 10104 (2010).
[52] H. C. Lin and Y. H. Lin, *Appl. Phys. Lett.* 98, 083503 (2011).
[53] Y. H. Lin, T. Y. Chu, W. L. Chu, Y. S. Tsou, Y. P. Chiu, F. Lu, W. C. Tsai, and S. T. Wu, *J. Nanomed. Nanotechnol.* S9, 001 (2011).
[54] Y. H. Lin, T. Y. Chu, Y. S. Tsou, K. H. Chang, and Y. P. Chiu, *Appl. Phys. Lett.* 101, 233502 (2012).
[55] Y. S. Tsou, K. H. Chang, and Y. H. Lin, *J. Appl. Phys.* 113, 244504 (2013).

In: Solar Power
Editor: Stephen Bailey

ISBN: 978-1-63321-317-3

Chapter 6

MODELLING OF ADVANCED PROMOTION MECHANISMS FOR SOLAR TECHNOLOGIES: NET-METERING AND CAPACITY AUCTIONING SCHEMES

Andreas Poullikkas[*]
Department of Electrical Engineering, Cyprus University of Technology, Limassol, Cyprus

ABSTRACT

In this work, an optimisation model for the integration of renewable energy sources for power generation (RES-E) technologies in power generation systems on a unit commitment basis is developed. The purpose of the optimization procedure is to account for RES-E net-metering scheme as well as for RES-E capacity auctions. The optimization model developed uses a genetic algorithm (GA) technique for the calculation of the required RES-E levy (or green tax) in the electricity bills. Also, this GA procedure enables the estimation of the level of the adequate (or eligible) feed-in-tariff to be offered to future RES-E systems, which do not participate in the net-metering and/or capacity auctioning procedure. In order to demonstrate the applicability of the optimisation procedure developed the case of rooftop residential photovoltaic systems are investigated for the net-metering option and

[*] Fax: +357.25.002635, E-mail: mspoul@ucy.ac.cy.

large commercial PV systems are examined in the case of capacity auctioning scheme. The results indicated that the required green tax, in order to promote the use of RES-E technologies, which is charged to the electricity customers through their electricity bills, is reduced with the reduction in the final auctioning price and the degree of net-metering penetration. This has a significant effect related to the reduction of electricity bills.

Keywords: Renewable energy sources; capacity auctioning; photovoltaics; rooftop PVs; feed in tariff; net metering

1. Introduction

Successful promotion financial mechanisms have been instrumental in encouraging investments and in stimulating the development of renewable energy sources for power generation (RES-E). Despite the extensive experience in policy design acquired over the past decade, the need to craft and implement innovative policies as well as learn from best practices remains important in addressing prevalent barriers to deployment [12]. The rapid worldwide expansion of RES-E technologies in recent years has been largely driven by support policies. Typically, these aim to address market failures in an effort to promote the uptake of RES-E technologies while achieving a number of other objectives, including energy diversification, the development of a local industry and job creation [6, 9]. Among the common policies implemented, tariff-based support schemes, such as the feed in tariffs (FiTs) and the feed-in premiums, have proven to be very effective in stimulating growth in RES-E technologies. Over the recent past, net-metering schemes and RES-E capacity auction schemes have gained popularity given their inherent advantage in minimising support costs and regulating the deployment of RES-E systems [2].

A FiT scheme provides a guaranteed premium price to the renewable electricity producer and put an obligation on the grid operators to purchase the generated electricity output. The price is typically guaranteed for a long period in order to encourage investment in new RES-E plants. FiT schemes are well known for their success in deploying large amounts of wind, biomass and solar energy (both photovoltaics (PV) and concentrated solar power systems) mainly in Germany, Denmark and Spain. The biggest advantage of FiT schemes is the long-term certainty of financial support, which lowers investment risks considerably. An overview of the FiT supporting schemes available in Europe is provided in [21].

Net metering is an electricity policy, which allows utility customers to offset some, or all of their electricity use with self produced electricity from RES-E systems [3]. Net metering works by utilizing a meter that is able to spin and record energy flow in both directions. The meter spins forward when a customer is drawing power from the utility grid (i.e., using more energy than they are producing) and spins backward when energy is being sent back to the grid (i.e., using less energy than they are producing). Another way is when each channel is metered separately and the one is subtracted from the other. In both cases at the end of a given month, the customer is billed only for the net electricity used. Net metering works only for grid connected systems and what makes it so beneficial, besides offsetting a home's energy consumption with a RES-E system, is that excess energy sent to the utility can be sold back at retail price. If more energy is produced than consumed, depending on the legislative arrangements in place as well as the arrangements with the electric utility, producers may receive benefit for this positive balance, such as, renewable energy credits (REC), which is credited on the customer's account toward the next billing cycle. If at the end of the year a surplus remains, then the customer depending on the utility policy may (a) paid for the total REC collected at avoidance cost rate or retail cost rate, or, (b) the total REC collected can be transferred and could be used as a compensation for a possible negative balance in the following years, or, (c) the total REC collected are granted back to the utility [14]. Net metering is gaining recognition as an effective RES-E promotion incentive and it is currently used in the USA and Australia. In Europe only Denmark and Italy are using net metering [21].

Some variations of the net metering mechanism are the time of use (TOU) metering and the market rate metering. TOU net metering employs a specialized reversible smartmeter that is programmed to determine electricity usage any time during the day. TOU allows utility rates and charges to be assessed based on when the electricity was used, i.e., day or night and seasonal rates. Typically the production cost of electricity is highest during the daytime peak usage period and low during the night, when usage is low. TOU metering is a significant issue for RES-E, since, for example, solar power systems tend to produce energy during the daytime peak-price period, and produce little or no power during the night period, when price is low. In market rate net metering systems, the user's energy use is priced dynamically according to some function of retail electric prices. The users' meters are programmed remotely to calculate the value and are read remotely. Net metering applies such variable pricing to excess power produced by qualifying systems [13].

RES-E capacity auctions are also known as "demand auctions" or "procurement auctions" or "contract for differences (CfD) FiTs (CfD-FiTs)", whereby the government issues a call for tenders to install a certain capacity of RES-E technology [11, 12, 15]. Project developers who participate in the auction submit a bid with a price per unit of electricity at which they are able to realise the project. The government evaluates the offers on the basis of the price and other criteria and signs a power purchasing agreement with the successful bidder. RES-E capacity auctions, despite some difficulties in implementation in the past, have become a popular policy tool in recent years. The number of countries that adopted RES-E capacity auctions increased from 9 in 2009 to at least 44 by early 2013, [12]. The renewed interest in auction schemes is driven by their potential to achieve deployment in a cost-efficient and regulated manner [8, 10]. Auction schemes have benefited from the rapidly decreasing costs of RES-E technologies, the increased number of project developers, their international exposure and know-how, and the considerable policy-design experience acquired over the last decade. When well designed, the price competition inherent to the auction scheme increases cost efficiency and allows price discovery of RES-E based electricity, avoiding potential windfall profits and underpayments. While auctions have become very attractive, they only benefit the successful bidders and tend to favour large players that are able to afford the associated administrative and transaction costs [1, 23].

Based on national energy plans as well as the size and maturity of the RES-E market, the design of auction schemes will reflect each country's priorities in terms of technology, volume and location [5]. Technology specific auctions allow for the promotion of certain technologies and diversification of the portfolio. In addition to selecting the technology, auctions can be site-specific [2]. The identification of sites with ideal resources and secured grid connection potentially reduces risks to investors. Technology neutral auctions have also served to promote RES-E technologies, which have even been able to compete with fossil fuels on certain occasions. The design of auctions allows governments to include other national priorities, the most common one being local content requirements [7].

In this work, an optimisation model for the integration of RES-E technologies in power generation systems on a unit commitment basis is developed. The purpose of the optimization procedure is to account for RES-E net-metering scheme as well as for RES-E capacity auctions. The optimization model developed uses a genetic algorithm (GA) technique for the calculation of the required RES-E levy (or green tax) in the electricity bills. Also, this GA procedure enables the estimation of the level of the adequate (or eligible) FiTs to be offered to future RES-E systems, which do not participate in the net-metering and/or capacity auctioning procedure. In order to demonstrate the applicability of the optimisation procedure developed (a) the case of rooftop residential photovoltaic systems are investigated for the net-metering option and (b) large commercial PV systems are examined in the case of capacity auctioning scheme. The applicability of the resulting optimization model is applied to the small isolated power generation system of the island of Cyprus.

In section 2, the benefits and misconceptions of net metering scheme are discussed; the simulation methodology used for this comparative assessment as well as the mathematical formulation developed for the net metering scheme computations are presented; and the different cases simulated are presented and the results obtained are discussed in detailed. In section 3, the Cyprus power system is briefly described; the modified GA and the simulation procedure are presented; the PV deployment integration analysis using the auctioning mechanism is set up and the results are discussed in detailed. The conclusions are summarized in section 4.

2. Net Metering Scheme for PV Residential Systems

There are benefits that accrue to the utility, the customer, and the community from net metering. For the utility, a well-designed net metering policy provides a simple, low cost, and easily administered way to deal with PV residential systems. Utilities obtain electricity and capacity from small, distributed PV installations. This is electricity they don't have to generate themselves or purchase on the market. For PV systems, this generation takes place every day of the year with a very high correlation with utility peak loads. Utilities call this a high load carrying capability since sunshine is relatively easy to predict. Thus, utilities obtain the benefit of additional capacity in their service territory paid for by their customers.

PV residential systems can, also, strengthen the distribution grid, especially in rural areas. This is because voltage tends to drop at the end of long distribution lines when loads are high, and if it drops below a threshold level, the breakers will trip and a temporary blackout occurs. Grid connected

PV systems tied to the distribution grid strengthen voltage and improve overall service. And this grid support can defer maintenance and upgrades in the power distribution system, which is a tangible benefit to utilities. Customers benefit from net metering of PV residential systems because they obtain a long-term guarantee of low utility bills. Communities benefit from the investment in local generation. This investment not only increases local property values but increases local business opportunities as well. It is the difference between paying rent and paying a mortgage [4].

There are also some misconceptions about net metering, such as that net metering hurts the utility bottom line by reducing revenues. This argument is similar to the one against energy efficiency that customers reducing their purchases of electricity hurt utility revenues. This would be true if all households bought a PV system and put it on their roofs. However, the current PV market is increasing, therefore, any net metering policy should receive regular review to monitor progress of the technology and development of the market [22]. If PVs, and especially energy efficiency, which has a much larger potential for impacting rates than PVs, gets to the point where it actually reduces utility revenues, then rates should be restructured to guarantee that service.

Another misconception is that net metering represents a subsidy from one group of customers to another. This argument has to do with the methodology that utilities use to charge customers. The argument is that utilities charge all customers in the same class a single rate, which represents an average cost of doing business plus profit. Thus a household who uses a lot of electricity during the day when the cost of obtaining electricity is higher pays the same as the household who uses electricity at night during off-peak hours. One could argue that one type of consumer subsidizes another based on patterns of consumption, etc. Utilities and their customers have supported this averaging formula for years. For example, building a new home represents a cost for a utility because it must invest in new generating capacity in order to supply this electricity. Therefore, customers subsidize solar systems through net metering no more than they subsidize construction of new homes. Both represent expanding business opportunities, and electric utilities have figured out a way to accommodate this economic growth through existing rate structures for more than a century [26]. Still complicated issues such as capacity payment as well contribution of net metering PV systems to the power system ancillary services need to be an in depth examination.

A final misconception is that net metering represents a burden for small utilities. The opposite is actually true because large organizations are better equipped to handle more complicated arrangements. Net metering requires only a net meter, no new rates to establish and no new procedures. All that is required is that the utility adds a line in the ledger for each net metering customer to carry forward credits until the end of the year. Compare this with the alternative of FiT supporting scheme, which requires installation of another meter. Then the utility must make special trips to read this meter and readjust its accounting procedures to keep track of another meter for a single account. A survey found that the cost of reading the extra meters for residential PV systems alone outweighed the cost of net metering [13].

In this section, a comparative assessment of net metering versus FiT supporting schemes for residential PV systems is carried out. For the analysis the IPP algorithm version 2.1 (independent power producer technology selection algorithm) software tool is employed [16]. In the case of a FiT supporting scheme, for the computation of the annual electricity unit cost as well as for the various financial indicators, in half hour intervals, the methodology presented in [19] and [20] is used. In the case of net metering supporting scheme the formulation is developed in this work and integrated in the IPP algorithm version 2.1 [16]. A typical household integrated with a rooftop PV system is investigated. The effect of the size of PV system with respect to the net metering supporting scheme is examined by varying the PV capacity from 1kWp up to 7kW and the effect of the electricity retail cost rate is investigated by varying the electricity retail cost rate from 16€c/kWh up to 24€c/kWh. For the purpose of this work, in order to demonstrate the applicability of the developed net metering formulation, available data from the Cyprus electricity market is used.

2.1. Mathematical Formulation

In order to carry out a comparative assessment of FiT and net metering RES-E supporting schemes for all simulations, the IPP algorithm version 2.1 software tool is employed [16]. The software, allows the user to input various technical, financial and environmental parameters of a PV system, such as efficiencies, solar potential, discount rates, FiTs, etc. Then the operation of the PV system is simulated and the key financial feasibility indicators, such as internal rate of return (IRR), payback period, net present value (NPV), etc., are calculated based on the following algorithm (a) calculate solar radiation in plane of PV system, (b) calculate electrical energy delivered by PV system, (c) calculate system losses, such as, temperature effects and inverter losses, (d) calculate electrical energy delivered to the grid, (e) calculate RES-E avoidance cost, (f) calculate CO_2 emissions avoided benefit and (g) calculate financial feasibility indicators assuming that the initial investment year is year 0, the costs and credits are given in year 0 terms, thus any inflation rate (or the escalation rate) is applied from year 1 onwards and the timing of cash flows occurs at the end of the year. During the simulations procedure the following financial feasibility indicators are calculated: (a) electricity unit cost before tax (in €/kWh), (b) after tax cash flow (in €), (c) after tax NPV, (d) after tax IRR and (e) after tax payback period [16].

In the case of a FiT scheme, for the computation of the annual electricity unit cost as well as for the various financial indicators, in half hour intervals, the methodology used is presented in [19] and [20]. In the case of net metering supporting scheme the following formulation is developed and integrated in the IPP algorithm version 2.1 [16].

The electricity demand of a household *E*, in kWh, is given by:

$$E = \sum_{j=1}^{m} \sum_{i=1}^{n} E_{ij} = \sum_{j=1}^{m} \sum_{i=1}^{n} P_{Dij}\, T_{ij}, \tag{1}$$

where P_D is the power demand in kWe during the specified time period T (i.e., for every 30 minute period, $T = 0.5$), $j = 1,2,3,\ldots,m$, is the number of day under consideration (i.e., for one year, $m = 365$) and $i = 1,2,3,\ldots,n$, is the number of the daily time period (i.e., for every 30 minutes, $n = 48$).

The electricity generation by the PV system *G*, in kWh, is given by:

$$G = \sum_{j=1}^{m} \sum_{i=1}^{n} G_{ij} = \sum_{j=1}^{m} \sum_{i=1}^{n} I_{ij}\, P_{PV} A_{PV} \eta_{PV} \eta_S, \tag{2}$$

where I is the available solar potential in kWh/m^2, P_{PV} is the installed capacity of the PV system in kWp, A_{PV} is the PV panels frame area in m^2/kWp, η_{PV} is the efficiency of the photovoltaic panels in % and η_S represents the additional losses (such as, inverter and wire losses) in %.

The net electricity based on the net metering reading *N*, in kWh, is simply the difference between the energy demand *E*, equation (1) and the generation by the PV system *G*, equation (2):

$$N = \sum_{j=1}^{m} \sum_{i=1}^{n} N_{ij} = \sum_{j=1}^{m} \sum_{i=1}^{n} E_{ij} - \sum_{j=1}^{m} \sum_{i=1}^{n} G_{ij}. \tag{3}$$

Taking into account a simple net metering scheme or more advance net metering schemes, such as, TOU or market rate net metering, the electricity demand cost C_E, in €, is given by:

$$C_E = \sum_{j=1}^{m} \sum_{i=1}^{n} C_{Eij} = \sum_{j=1}^{m} \sum_{i=1}^{n} E_{ij} Z_{Rij}, \quad (4)$$

where Z_R is the electricity retail cost rate in €/kWh. The net metering electricity cost C_{NET}, in €, is given by:

$$C_{NET} = \sum_{j=1}^{m} \sum_{i=1}^{n} C_{NETij} = \sum_{j=1}^{m} \sum_{i=1}^{n} N_{ij} Z_{Rij}. \quad (5)$$

The net savings S_{NET}, in €, is given by:

$$S_{NET} = \sum_{j=1}^{m} \sum_{i=1}^{n} S_{NETij} = \sum_{j=1}^{m} \sum_{i=1}^{n} \left(C_{Eij} - C_{NETij}\right). \quad (6)$$

The renewable energy credits *REC*, in kWh, for a particular month *k* (where $k = 1,2,3, \dots ,12$) are provided by the relation:

$$\text{if}\left\{N_k = \left[\sum_{j=1}^{m} \sum_{i=1}^{n} N_{ij}\right]_k < 0\right\}, \quad (7)$$

then, $REC_k = \text{abs}\,(N_k)$.

Finally, the benefit based on the total REC generated over a year C_{REC}, in €, is given by:

$$C_{REC} = \sum_{k=1}^{12} REC_k Z_{Ak}. \quad (8)$$

where Z_A is the avoidance cost rate in €/kWh for a particular month *k*. In the case the benefit is paid for the total *REC* collected at retail cost rate, then Z_A can be substitute with Z_R in equation (8) above. Finally, if the total REC collected are granted back to the utility then $C_{REC} = 0$.

2.2. Case Study

For the comparative assessment of net metering and FiT RES-E supporting schemes a typical household is examined. In order take into account and compare all available net metering REC policies the following scenarios are investigated for a household integrated with a rooftop PV system: (a) FiT supporting scheme only, (b) FiT supporting scheme and taking into account household electricity consumption (i.e., monthly electricity bill of the household), (c) Net metering supporting scheme with REC granted back to the utility, (d) Net metering supporting scheme with REC payment at avoidance cost rate and (e) Net metering supporting scheme with REC payment at retail cost rate.

2.2.1. Data and Assumptions

The input data and assumptions used for the analysis are tabulated in Table 1. For the purpose of this comparative study we assume the installation and operation of a rooftop PV system on a typical household with an annual power demand as illustrated in Figure 1 and of a total annual electricity demand of 10310kWh. The monthly power demand is presented Figure 2 with a maximum power demand of 2.6kW and a minimum power demand of 0.6kW. The household is assumed to be located in an area with available annual solar potential of 1968kWh/m^2. The available annual solar potential in 30 minutes intervals used during the simulations is presented in Figure 3. A typical mono-Si solar PV module has been selected with a capacity of 185W, efficiency 14.2% and area of 1.3m^2.

The effect of the size of PV system with respect to the net metering supporting scheme is examined in this analysis by varying the PV capacity from 1kWp, in steps of 1kWp, up to 7kWp, which is the maximum allowable installed capacity for rooftop PV systems in Cyprus. Thus for 1kWp PV system approximately 6 modules are required with a total required area of approximately 7m^2/kW and for 7kWp PV system approximately 38 modules are required with a total required area of approximately 50m^2/kW.

For the PV system a capital expenditure of 2700€/kW is used and for the annual operation and maintenance a typical value of 3€/kW-month is assumed [20]. For all simulations, costs are updated to 2012 values. In particular, the economic life of the PV system is assumed for 20 years with a discount rate of 8%, an average annual inflation rate of 2.3% and an average loan interest of 7.1%.

In order to calculate the after-tax cash flows and after-tax financial indicators a single 10% income tax rate is assumed, that remains constant throughout the PV system life and applied to net income. In the case of FiT supporting scheme a FiT of 28€c/kWh is assumed.

Table 1. Summary of input data and assumptions

Parameter	Value
Technical data	
Household annual power demand	see Figure 1 for annual 30 minutes distribution power demand and Figure 2 for monthly distribution
Household annual electricity demand	10310kWh
Annual solar potential	1968kWh/m^2 (see Figure 3 for annual solar potential 30 minutes distribution)
PV module type	mono-Si
PV module area	7.03m^2/kW
PV module efficiency	14.20%
PV system losses	18.04%
PV capacity	1kWp -7kWp
Economic data	
PV specific capital cost	2700€/kW
PV annual operation and maintenance cost	3€/kW-month
PV system economic life	20 years
Discount rate	8.0%
Loan interest	7.8%
Annual inflation	2.3%
Annual income tax rate	10%
FiT supporting scheme	28€c/kWh
Electricity retail cost rate	16€c/kWh - 24€c/kWh
Avoidance cost rate	10.67€c/kWh - 16.00€c/kWh (see Figure 4)
Environmental data	
CO_2 indicator	800g/kWh

The effect of the electricity retail cost rate and the effect of the avoidance cost rate to the net metering supporting scheme is investigated by varying the electricity retail cost rate from 16€c/kWh, in steps of 2€c/kWh, up to 24€c/kWh. The avoidance cost rate is then changing accordingly as illustrated in Figure 4.

To estimate the annual CO_2 avoided emissions from the operation of the PV system, a CO_2 environmental indicator of 800g/kWh is used.

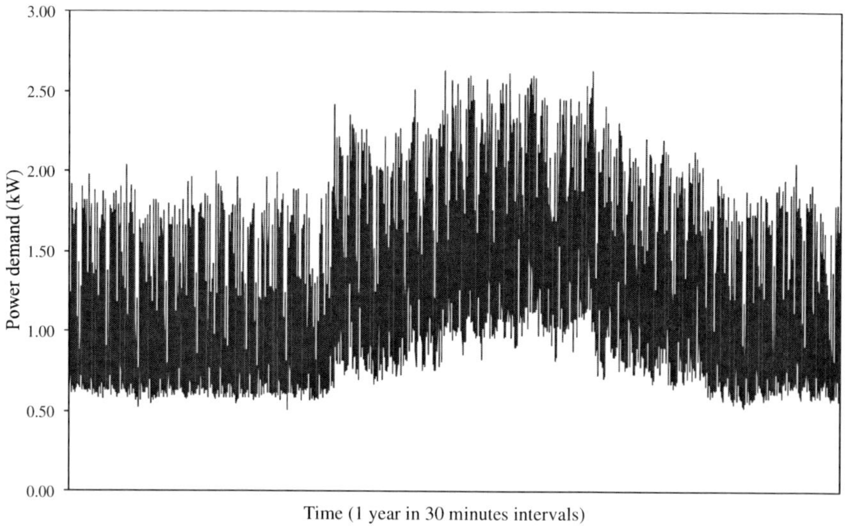

Figure 1. Annual power demand.

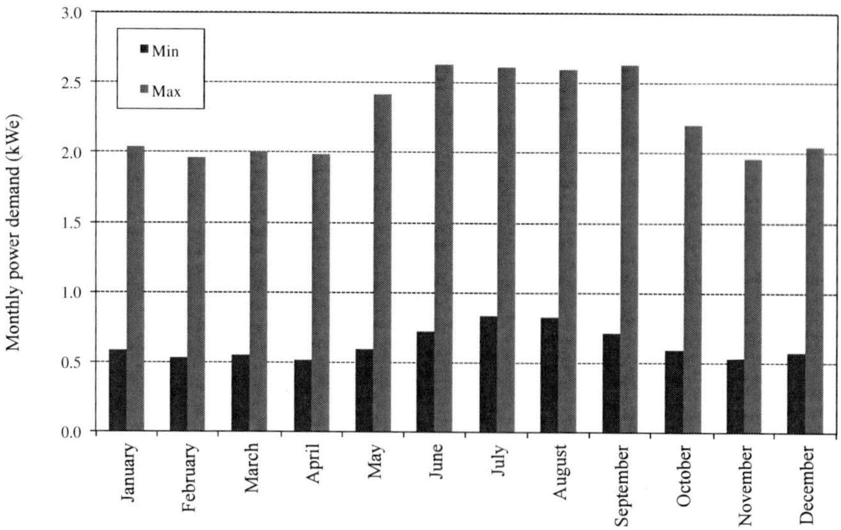

Figure 2. Monthly power demand.

2.2.2. Simulations Results and Discussion

The results concerning the household annual electricity demand and the electricity generation from the capacity range of rooftop PV systems investigated, as well as, the results concerning net metering reading and RECs credited are tabulated in Table 2.

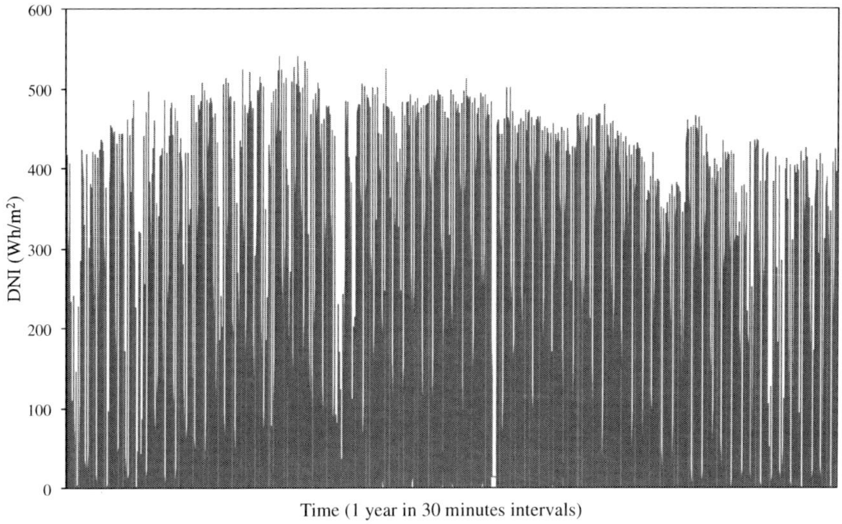

Figure 3. Annual solar potential.

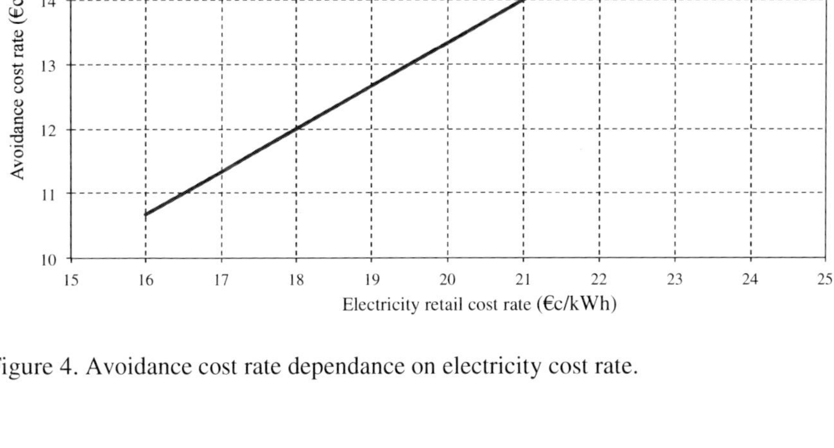

Figure 4. Avoidance cost rate dependance on electricity cost rate.

Table 2. Simulations annual results

Parameter	Unit	PV system capacity						
		1kWp	2kWp	3kWp	4kWp	5kWp	6kWp	7kWp
Electricity demand	kWh	10310	10310	10310	10310	10310	10310	10310
PV electricity generation	kWh	1611	3221	4832	6442	8053	9664	11274
Net metering reading	kWh	8699	7088	5478	3867	2257	646	-965
RECs	-	0	0	0	0	64	361	1126
Annual CO_2 avoided emissions	t_{CO2}	1	3	4	5	6	8	9
Annual barrels of crude oil not consumed	bbl	3	5	8	11	13	16	19

We observe that net metering reading decreases with the increase in the PV system capacity. In the case of 7kWp PV system the net metering reading is negative meaning that a surplus, of electricity generated from the PV system, of 965kWh exists. However, the credited RECs are 1126 since the clearance of surplus electricity is carried out for each month. Although in the case of 5kWp and 6kWp PV systems there is no annual surplus of the generated electricity the credited RECs are 64 and 361 respectively.

The monthly energy balance in the case of 3kWp, 5kWp and 7kWp PV system are illustrated in Figure 5-Figure 7 respectively. In the case of 3kWp PV system, shown in Figure 5, the results indicated that for all months there is no surplus of electricity generated from the PV systems. However, in the case of 5kWp system, shown in Figure 6 the results indicated that in April a total of 64kWh of electricity is exported to the grid resulting to an equivalent number of RECs benefit. For a PV system of 7kWp, shown in Figure 7, electricity is exported to the grid for seven months resulting to a number of RECs benefit of 1126. Daily results for April concerning various PV system capacities are presented in Figure 8. In the case of 1kWp no electricity is exported to the grid, however, in the case of using 5kWp or 7kWp PV systems electricity is exported to the grid.

Referring to Table 2, the annual avoided CO_2 emissions are estimated at $1t_{CO2}$, which corresponds to 3 barrels of crude oil not consumed in the case of 1kWp PV system. In the case of 7kWp PV system the annual avoided CO_2

emissions are $9t_{CO2}$, which corresponds to 19 barrels of crude oil not consumed.

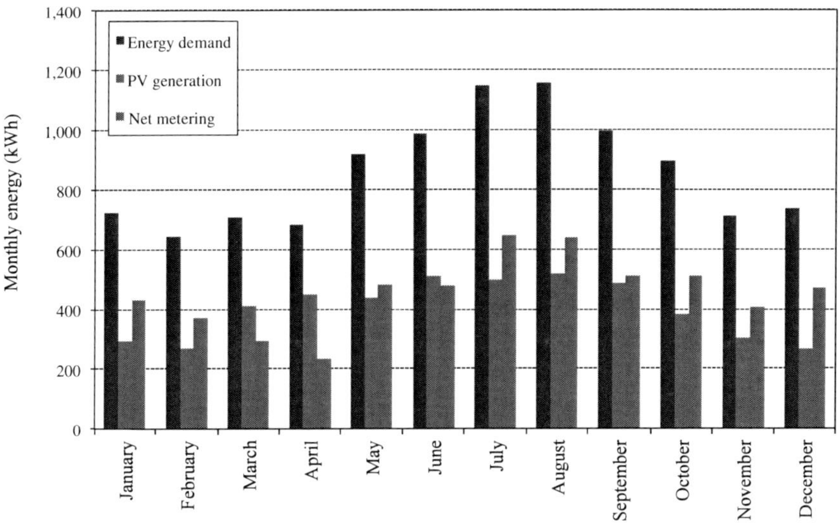

Figure 5. Monthly energy balance in the case of 3kWp PV installation.

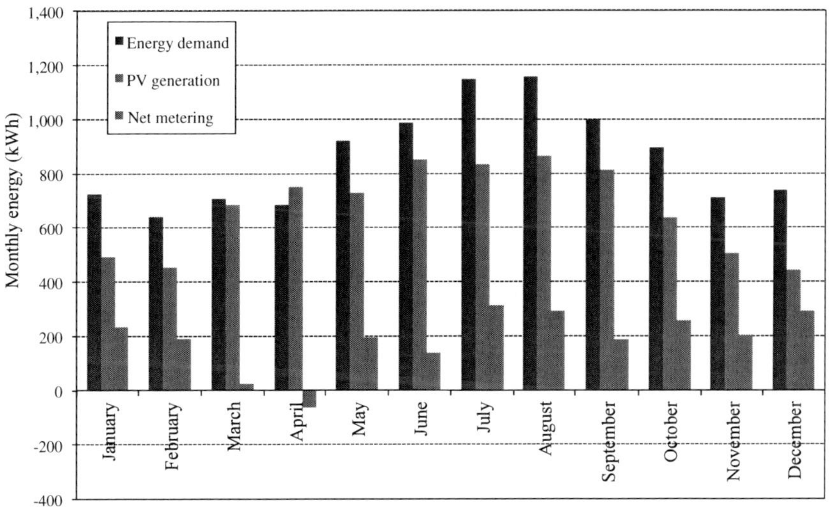

Figure 6. Monthly energy balance in the case of 5kWp PV installation.

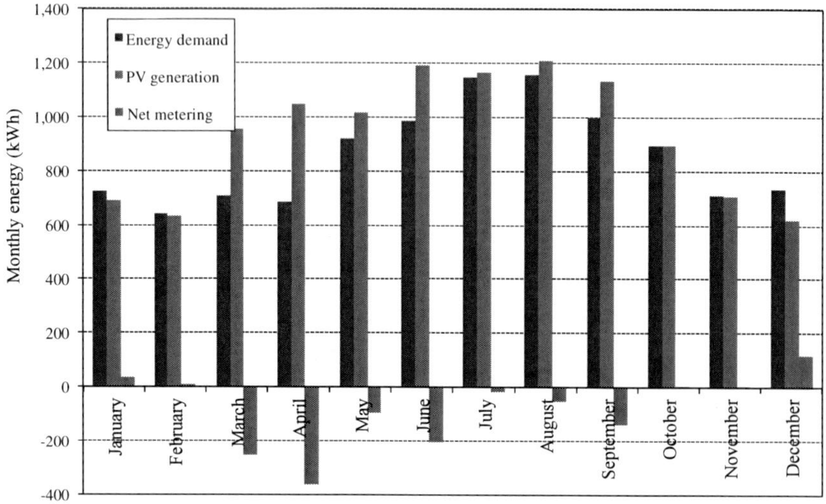

Figure 7. Monthly energy balance in the case of 7kWp PV installation.

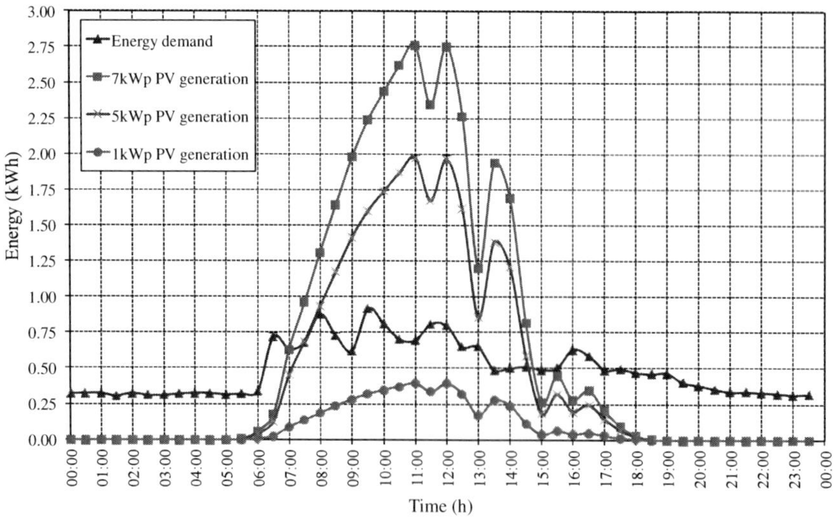

Figure 8. Energy demand and PV generation in April.

The comparative results concerning after tax NPV for the various scenarios examined are presented in Figure 9. We observe that in the case of evaluating the economics of the PV system under FiT supporting scheme without taking into account the household electricity bill is profitable with an after tax NPV ranging from €909 to €6361 depending on the capacity of the installed rooftop PV system. In the case, however, of residential PV system installations except from the FiT support a utility customer is required to take into account the cost of electricity imported for satisfying the energy needs of the household as illustrated by the bottom lines of Figure 9. We observe that depending on the level of electricity retail cost rate and the rooftop PV system capacity negative after tax NPVs ranging from €8215 to €20955 are calculated.

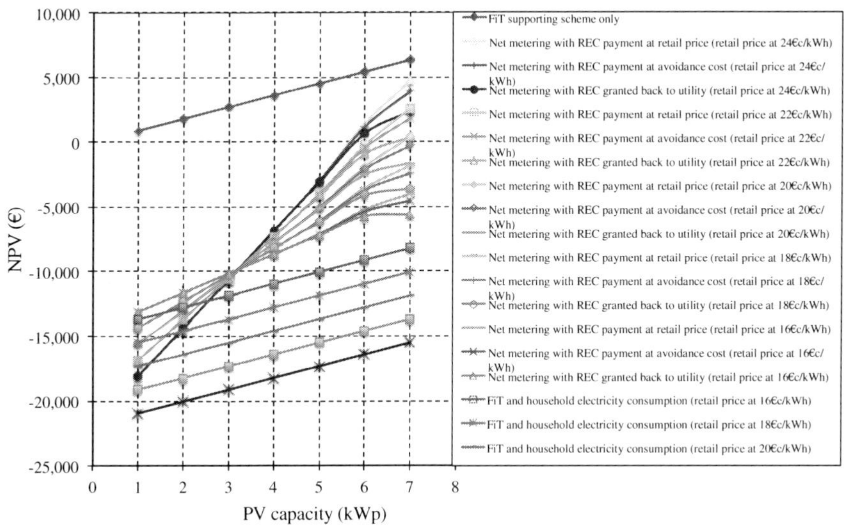

Figure 9. After tax NPV results.

In the case of net metering scenarios examined all computations provided after tax NPV results greater than the above scenario (i.e., FiT and household electricity consumption). Also, depending on the rooftop PV system capacity, the electricity retail cost rate and the REC policy examined indicated that after tax NPV becomes positive reaching a value of €4763.

The comparative results concerning after tax IRR are illustrated in Figure 10. The results of the scenario FiT and household electricity consumption are not illustrated since IRR values calculated are below 0%. Based on this analysis, using data from the Cyprus electricity market, it is clear that net metering supporting scheme becomes profitable for values of after tax IRR above 8%. This is the case when electricity retail cost price is above 22€c/kWh.

3. Capacity Auctioning Scheme for Large PV Systems

In a previous work [19], an optimization model for the integration of RES-E technologies in power generation systems on a unit commitment basis was developed. The purpose of the optimization procedure was to assess the unavoidable increase (or decrease) in the cost of electricity of a given power generation system at different RES-E penetration levels. The optimization model developed uses a genetic algorithm (GA) technique for the calculation of both the additional cost (or benefit) of electricity due to the large penetration of RES-E technologies as well as the required RES-E levy (or green tax) in the electricity bills in order to fund this RES-E penetration. The algorithm combines the WASP IV software [25], for optimal expansion plan for a given power generating system and the IPP v2.1 software [16] for the optimum cost of electricity produced from both conventional and RES-E technologies. The GA procedure enables the estimation of the level of the adequate (or eligible) FiTs to be offered to future RES-E systems. The applicability of the developed optimization model was applied to the small isolated power generation system of the island of Cyprus.

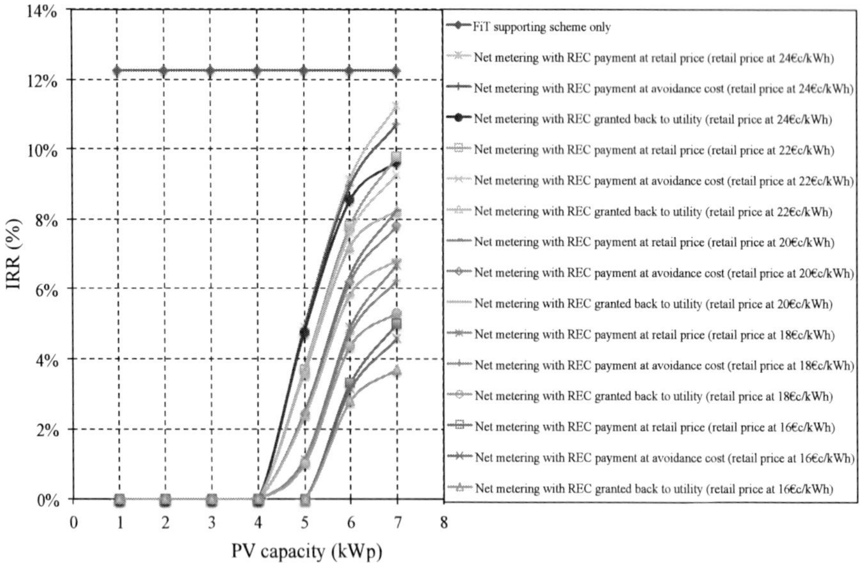

Figure 10. After tax IRR results.

In this section, a modification of the above GA is developed in order to extent the applicability of the optimization procedure in [19] to account for RES-E capacity auctions. In particular, relevent cost functions are integrated within the GA for solar PV capacity electricity prices. The applicability of the resulting optimization model is applied, as previously in [19], to the small isolated power generation system of the island of Cyprus.

3.1. Description of Cyprus Power Generation System

Cyprus is an island with no indigenous hydrocarbon energy sources. This means that its power generation system operates in isolation and totally relies on imported fuels for electricity generation. Currently, the primary imported fuel used in electricity generation is heavy fuel oil (HFO) and gasoil. Cyprus power generation system consists of three thermal power stations with a total installed capacity of 1480MWe. Moni power station is located on the south coast of Cyprus, to the east of Limassol and consists of 4x37.5MWe gas turbines. Dhekelia power station is located on the southeast coast of Cyprus, to the east of Larnaca and consists of 6x60MWe steam turbines and two 51MWe internal combustion engines blocks. Finally, Vasilikos power station is the most recent power station located on the south coast between Limassol and Larnaca and consists of 3x130MWe steam turbines, 2x220MWe combined cycle technology units and a 38MWe gas turbine.

The steam units at Vasilikos are used for base load generation, while the steam units of Dhekelia are used for intermediate load generation. The gas turbines are mainly used during system peak loading. All stations use HFO for the steam turbine units and gasoil for the gas turbine units. The combined cycle units use gasoil as fuel until the arrival of natural gas in Cyprus, which is expected to be available on the island in 2016. Also, in 2016 the three steam turbines at Vasilikos power station will shift from the use of HFO to the use of natural gas. All three power stations are owned and operated by the Electricity Authority of Cyprus (EAC), which currently is the sole producer of electricity from conventional fuel on the island. This is despite the fact that the energy market in Cyprus has been liberalized since 2004 with the establishment of the Cyprus Energy Regulatory Authority (CERA) and the Cyprus Transmission System Operator (TSO).

Concerning the share of RES-E in electrical energy generation, it currently amounts to 147MW wind parks, a number of small-rooftop PV systems and PV parks with a capacity of up to 150kW, with a total combined grid connected capacity of approximately 30MWe, and to biomass gasification units (the majority of which use animal or domestic waste) with a total grid connected capacity of 10MWe. Recently the net metering scheme have been introduced in the case of residential PV systems with a capacity less than 3kW. Lastly a competitive auctioning process has been introduced in the case of commercial PV systems in order to reduce electricity selling prices.

3.2. Simulation Methodology

The optimization model developed [19], uses a GA technique for the calculation of both the electricity cost (or benefit) due to the integration of RES-E technologies as well as the required RES-E levy (or green tax) in the electricity bills in order to promote the use of RES-E technologies. A schematic diagram of the optimization flow chart is shown in Figure 11. The algorithm combines the WASP IV software [25], for optimal expansion plan for a given power generating system and the IPP v2.1 software [16] for the optimum cost of electricity produced from both conventional and RES-E technologies. Both have been used extensively during the past years for similar studies (e.g., [17, 19]). Recent expansion of the optimization model concerns the use of net-metering mechanism and the use of pumped energy storage systems [18].

In this work, a modification of the above GA is developed in order to extent the applicability of the optimization procedure to account for RES-E capacity auctions. In particular, cost functions are integrated within the GA for PV capacity electricity prices. A brief description of the main simulation software follows.

3.2.1. WASP IV Software

The future power generation system of Cyprus power industry is simulated using the Wien Automatic System Planning IV (WASP IV) software package [25], which is widely used for automatic generation planning. The WASP IV software package finds the optimal expansion plan for a given power generating system over a period of up to 30 years. The foreseen seasonal load duration curves, the efficiency, the maintenance period and the forced outage rate of each generating plant are taken into account. The objective function, which shows the overall cost of the generation system (existing and candidate generating plants), is composed of several components.

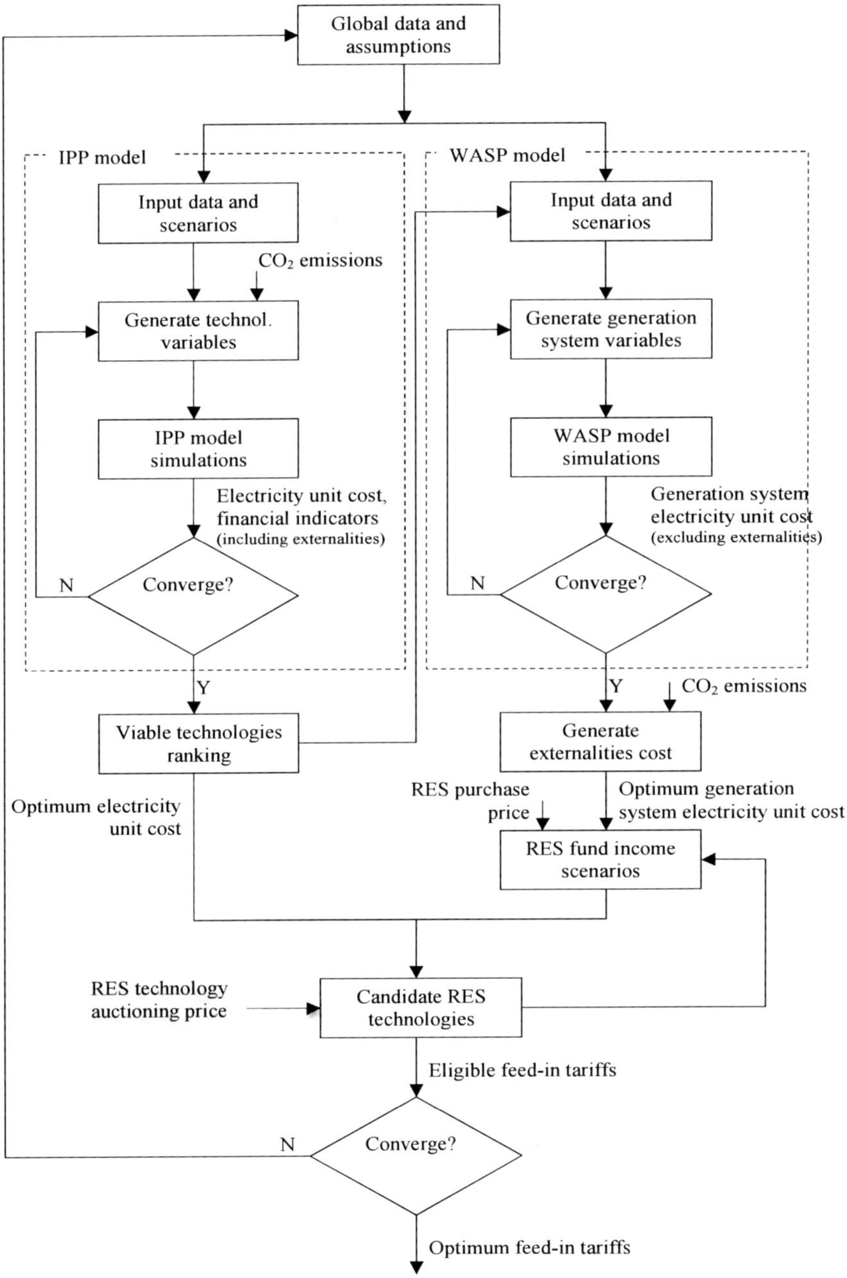

Figure 11. Flow chart of the optimization model.

The components, related to the candidate generating units, are the capital cost and the salvage capital cost. The components, which are related to both the existing and candidate generating units are the fuel cost, the fixed operation and maintenance costs, such as, staff cost, insurance charges, rates and fixed maintenance, the variable operation and maintenance costs, such as, spare parts, chemicals, oils, consumables, town water and sewage. The cost to the national economy of the energy not served (ENS) because of shortage of capacity or interruptions is, also, taken into consideration.

The WASP package was originally developed in the United States for the needs of the International Atomic Energy Agency (IAEA). It is the most frequently used and best-proven program for electric capacity expansion analysis. It is used for long-term expansion planning for a period of up to 30 years and compares the total costs for the whole generation system for a number of candidate units. In the production simulation of WASP, a one-year period is divided into, at most, 12 sub-periods for each of which probabilistic simulation is applied. Equivalent load duration curves in the probabilistic simulation are approximated using Fourier series. The Fourier expansion makes it computationally simple to convolve and deconvolve generating units in the probabilistic simulation. The decision of the optimum expansion plan is made by the use of forward dynamic programming. The number of units for each candidate plant type that may be selected each year, in addition to other practical factors that may constrain the solution is specified. If the solution is limited by any such constraints, the input parameters can be adjusted and the model re-run. The dynamic programming optimization is repeated until the optimum solution is found. Each possible sequence of power units added to the system (expansion plan) meeting the constraints is evaluated by means of a cost function (the objective function), which is composed of (a) capital investment costs, I, (b) salvage value of investment costs, S, (c) fuel costs, F, (d) non-fuel operation and maintenance costs, M, and (e) cost of energy not served, Φ.

Thus,

$$B_j = \sum_{j=1}^{T} \left(I_{jt} - S_{jt} + F_{jt} + M_{jt} + \Phi_{jt} \right), \tag{9}$$

where, B_j is the objective function attached to the expansion plan j, t is the time in years (1, 2, …., T) and T is the length of the study period (total number of years). All costs are discounted to a reference date at a given discount rate. The optimum expansion plan is the $\min B_j$ among all j. Details of the optimization algorithm implementing the above mathematical formulation can be found in [19].

3.2.2. IPP v2.1 Software

In order to calculate the cost of electricity from the various RES-E candidate technologies each plant operation is simulated using the IPP v2.1 software [16]. The software emerged from a continued research and development in the field of software development for the needs of power industry. This user-friendly software tool can be used for the selection of an appropriate least cost power generation technology in competitive electricity markets. The software takes into account the capital cost, the fuel consumption and cost, the operation cost, the maintenance cost, the plant load factor, etc. All costs are discounted to a reference date at a given discount rate. Each run can handle 50 different candidate schemes simultaneously. Based on the above input parameters for each candidate technology the algorithm calculates the least cost power generation configuration in real prices and the ranking order of the candidate schemes. A brief description of the optimization procedure is given below.

The technical and economic parameters of each candidate power generation technology are taken into account based on the cost function:

$$\min\left(\frac{\partial c}{\partial k}\right) = \min\left\{\frac{\sum_{j=0}^{N}\left[\frac{\left(\frac{\partial C_{Cj}}{\partial k}+\frac{\partial C_{Fj}}{\partial k}+\frac{\partial C_{OMFj}}{\partial k}+\frac{\partial C_{OMVj}}{\partial k}\right)}{(1+i)^{j}}\right]}{\sum_{j=0}^{N}\left[\frac{\frac{\partial P_j}{\partial k}}{(1+i)^{j}}\right]}\right\}, \tag{10}$$

where c is the final cost of electricity in €/kWh, in real prices, for the candidate technology k, C_{Cj} is the capital cost function in €, C_{Fj} is the fuel cost function in €, C_{OMFj} is the fixed O&M cost function in €, C_{OMVj} is the variable O&M cost function in €, P_j is the total electricity production in kWh, j=1,2,...N is the periods (e.g., years) of installation and operation of the power generation technology and i is the discount rate. The least cost solution is calculated by:

$$\text{least cost solution} = \min\left(\frac{\partial c}{\partial k}\right). \tag{11}$$

During the simulations procedure the following financial feasibility indicators are calculated, such as, electricity unit cost or benefit before tax (in €/kWh), after tax cash flow (in €), after tax NPV (net present value: the value of all future cash flows, discounted at the discount rate, in today's currency), after tax IRR (internal rate of return: the discount rate that causes the NPV of the project to be zero and is calculated using the after tax cash flows. Note that the IRR is undefined in certain cases, notably if the project yields immediate positive cash flow in year zero) and after tax PBP (payback period: the number of years it takes for the cash flow, excluding debt payments, to equal the total investment which is equal to the sum of the debt and equity). Details of the optimization algorithm implementing the above mathematical formulation can be found in [16].

3.3. PV Deployment Integration Analysis Using Auctioning Mechanism

In this section, we demonstrate the applicability of the optimization procedure develop in [19], for the optimum estimation of RES-E required FiT, to account for RES-E capacity auctions. In order to demonstrate the applicability of the modified optimization procedure presented in section 3 the case of PV capacity auctioning for commercial systems with a capacity greater than 150kW is examined.

3.3.1. Data and Assumptions

The forecasted annual peak load demand and the forecasted annual electrical energy for the period from 2014 to 2020 are presented in Figure 12. The fuel costs for HFO, gasoil and natural gas for the period under investigation were based on the forecast scenario illustrated in Figure 13.

For the purposes of this study the technical and economic parameters for each of the existing and committed conventional power generation units of the Cyprus power system are taken into account. All data is based on actual costs derived from manpower, spares requirements, maintenance costs, unit availabilities and operation efficiencies. The commissioning and retirement years of the existing units are also taken into account and follow the long-term strategic planning that has been developed for the Cyprus generation system [19]. Also, the technical and economic parameters for each of the existing and committed RES-E units already in operation are taken into account. A summary of the existing and committed RES-E units taken into account in this study is tabulated in Table 3.

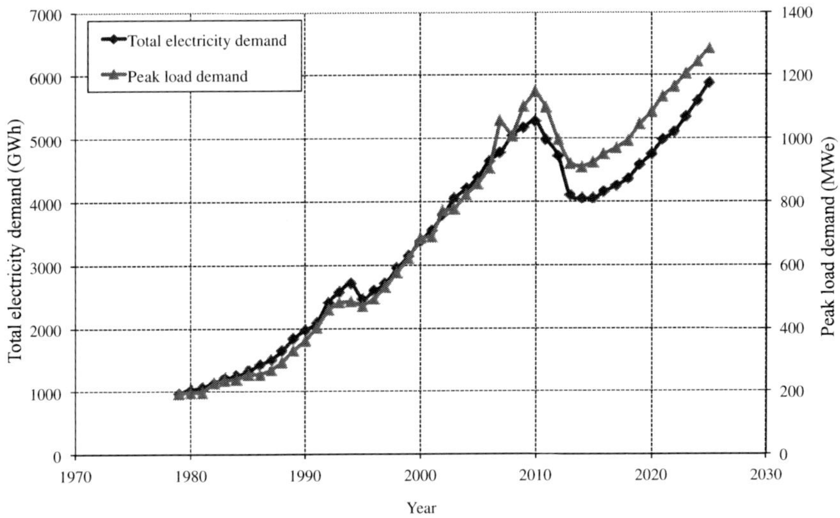

Figure 12. Energy and peak load demand for the period 1979 – 2025.

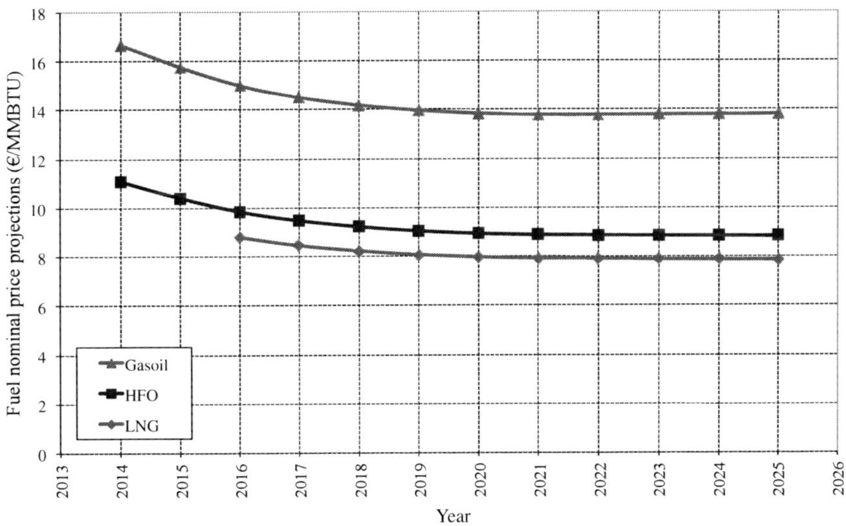

Figure 13. Fuel cost projections (nominal values).

Table 3. Existing and committed RES-E units

Technology	Capacity (MWe)
Wind parks	147
Photovoltaic systems	30
Biomass systems	10

Based on Cyprus' national plant [19], one conventional power generation technology and four promising RES-E technologies are considered for the capacity expansion of the power generation system. These include (a) the natural gas combined cycle technology, (b) the wind technology, (c) the PV technology, (d) the biomass technology and (e) the CSP technology with 6 hours thermal storage. Natural gas combined cycle technology is considered as the business as usual (BAU) technology for the expansion of Cyprus power system. Concerning the RES-E technologies with the exception of the CSP technology, the rest of the technologies are assumed to grow at a steady rate, up to the year 2020. By the year 2020 the RES-E energy share is expected to reach 17.4% of total expected electricity demand as illustrated in Figure 14. Concerning the CSP technology, a single 25MWe parabolic trough CSP technology with thermal storage of 6 hours is assumed to become operational after 2015 and remain at this capacity up to the year 2020 [17]. All simulations concerning RES-E technologies are based on the following two parameters: (a) RES-E capacity installation priority (1. biomass, 2. wind, 3. PV and 4. CSP with 6h thermal storage) and (b) RES-E contribution to the power generation system capacity reserve margin (wind 0% contribution, PVs 30% contribution [24], biomass 100% contribution and CSP with 6h thermal storage 100% contribution).

The projected capital costs for the candidate RES-E technologies for the study period were based on 2013 real values The capital expenditure may include, among others, the following costs: (a) feasibility study, (b) land leasing, (c) power block and/or RES-E system, (d) development and design study, (e) license fees, (f) engineering works, (g) road construction, (h) connection transmission line, (i) substation, (j) commissioning and training, (k) contingencies, etc. For both optimization software package simulations all costs are updated to 2013 values. In particular, the horizon of this study covers a period of 7 years from 2014 up to 2020 with a discount rate of 8%, an average annual inflation rate of 2% and an average loan interest of 6%. In order to calculate the after-tax cash flows and after-tax financial indicators a single 12.5% income tax rate is assumed, that remains constant throughout the project life and applied to net income.

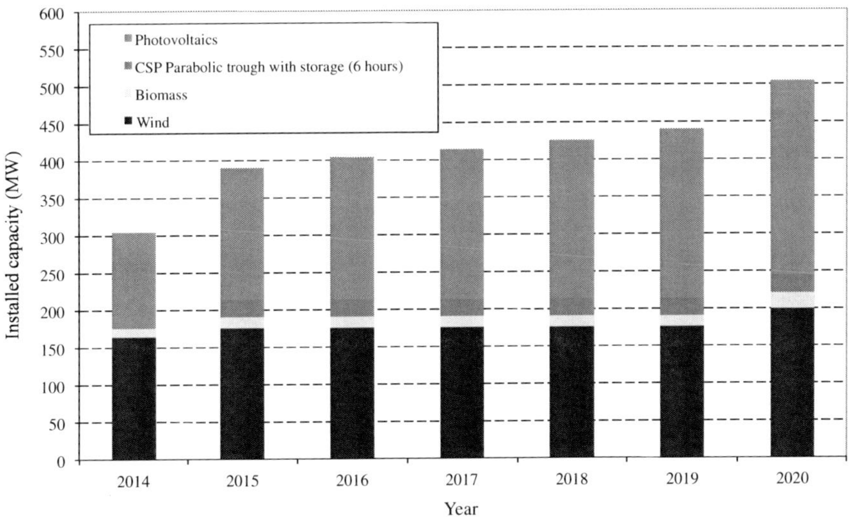

Figure 14. RES-E technologies expected installed capacity.

3.3.2. Simulations Results and Discussion

The results concerning the optimum power generation system energy production by the year 2020, with a RES-E energy share of 17.4% of total expected electricity demand, are presented in Figure 15. It is clear that the decreased energy generation from the conventional power generation plants is compensated by electricity produced from the RES-E technologies. The results concerning the optimum power generation system capacity reserve margin for the period from 2014 up to 2020 for each individual scenario are shown in Figure 16. It is observed that, in both scenarios examined, the capacity reserve margin of the Cyprus power generation system remains within the lower and upper limits of 20% and 40% respectively. This guarantees that the safety and reliability of electricity supply in the island can be maintained.

As mentioned earlier in this work we extend the applicability of the optimization procedure develop in [19], for the optimum estimation of RES-E required FiT, to account for RES-E capacity auctions. In order to demonstrate the applicability of the modified optimization procedure presented in section 3 the case of PV capacity auctioning for commercial systems with a capacity greater than 150kW is examined. Based on the optimization procedure (see

Figure 11), first the optimum FiTs for the RES-E technologies not participating in the capacity auctioning scheme are calculated.

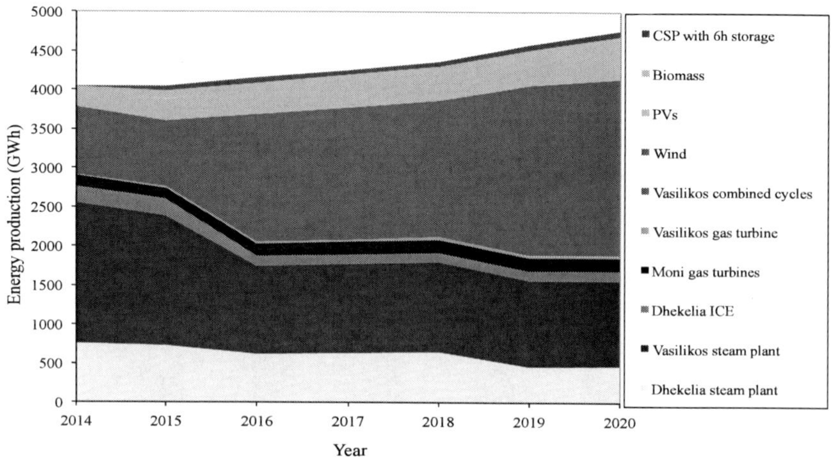

Figure 15. Power generation system energy production results for the scenario with natural gas after 2016.

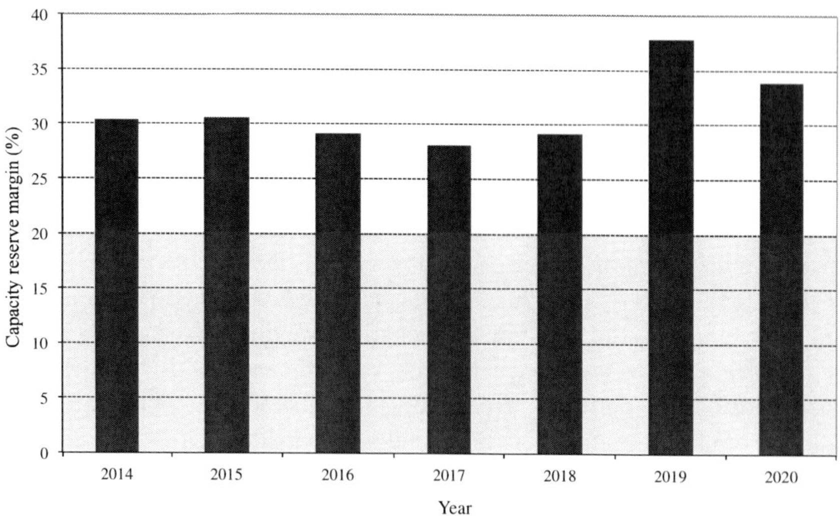

Figure 16. Power generation system capacity reserve margin results for the scenario with natural gas after 2016.

The results are tabulated in Table 4. In the case of residential PV systems with capacity of less than 3kW no optimum FiT is calculated since based on the current legal framework is Cyprus such category falls under the net-metering scheme and no FiT is required for the promotion of such systems. Secondly, the required RES fund income is calculated and the required RES-E levy (or green tax) to be imposed on the the electricity bills is evaluated. By using the capacity auctioning scheme the green tax level is expected to decrease with the final auctioning prices.

Table 4. Calculated optimum FiTs

Technology	FiT (€c/kWh)
Biomass systems	12.00
Wind technologies	14.56
CSP technologies (6 hours thermal storage)	26.54
Residential PV systems (capacity less than 3kW)	Net-metering scheme
Small commercial PV systems (capacity less than 150kW)	13.65
Commercial PV systems (capacity greater than 150kW)	Capacity auctioning scheme

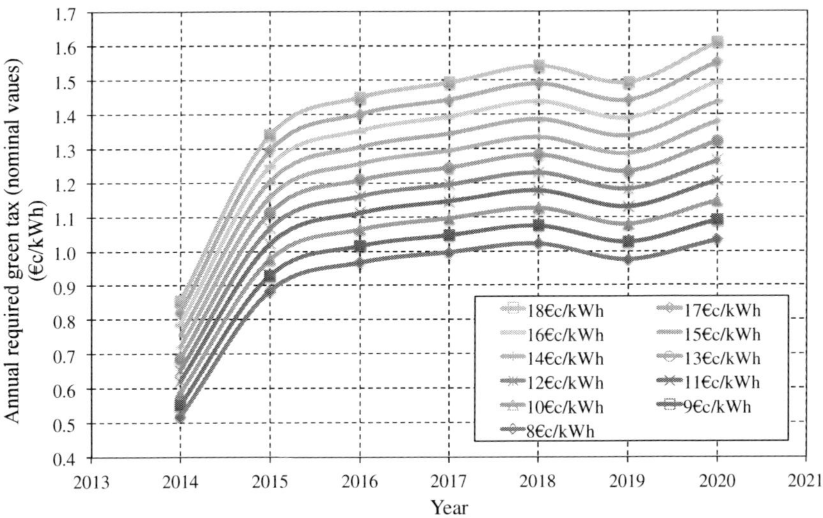

Figure 17. Annual required green tax for different PV auctioning price levels.

In order to demonstrate the above, a range of auctioning prices from 9€c/kWh to 18€c/kWh are examined in the case of PV capacity auctioning for commercial systems with a capacity greater than 150kW. The results concerning the required annual green tax are illustrated in Figure 17. As expected we observe that the required annual green tax is increased with the increase in RES-E penetration as well as with the increase in the final auctioning price level.

The average green tax for different auctioning price levels for the period 2014-2020 is presented in Figure 18 in which the green tax requirements are decreased with the decrease in the final auctioning electricity price. For example if the maximum auctioning price level for participation in the auctioning procedure is 18€c/kWh and the final auctioning price achieved after the competitive auctioning procedure is 10€c/kWh then the total decrease in the green tax for the period 2014-2020 will be €111,600,459 (in real prices). Recalling that the required green tax, in order to promote the use of RES-E technologies, is charged to the electricity customers through their electricity bills this is a significant amount resulting to reduce electricity bills.

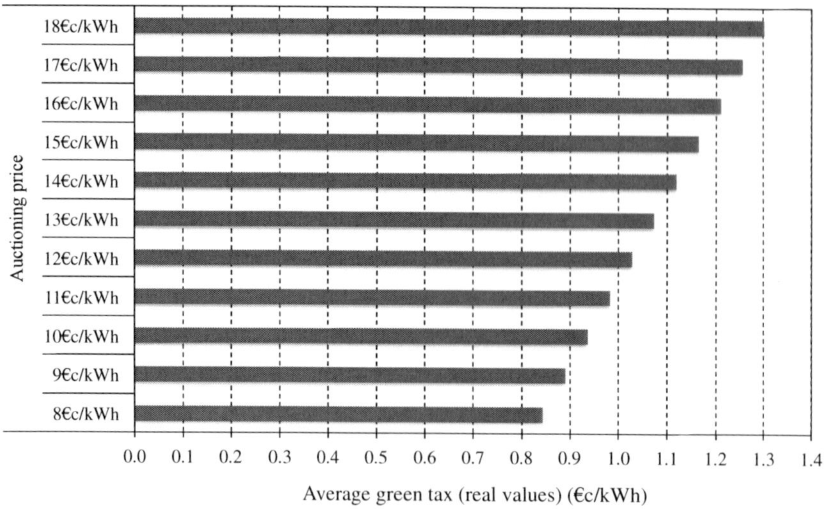

Figure 18. Average green tax for different PV auctioning price levels (period 2014-2020).

CONCLUSION

In this work, an optimisation model for the integration of RES-E technologies in power generation systems on a unit commitment basis was developed. The purpose of the optimization procedure was to account for RES-E net-metering scheme as well as for RES-E capacity auctions. The optimization model developed uses a GA technique for the calculation of the required RES-E levy (or green tax) in the electricity bills. Also, this GA procedure enables the estimation of the level of the adequate (or eligible) feed-in-tariff to be offered to future RES-E systems, which do not participate in the net-metering and/or capacity auctioning procedure. In order to demonstrate the applicability of the optimisation procedure developed the case of rooftop residential photovoltaic systems are investigated for the net-metering option and large commercial PV systems are examined in the case of capacity auctioning scheme. The results indicated that the required green tax, in order to promote the use of RES-E technologies, which is charged to the electricity customers through their electricity bills, is reduced with the reduction in the final auctioning price and the degree of net-metering penetration. This has a significant effect related to the reduction of electricity bills.

REFERENCES

[1] Boute A.A., 2012, "Promoting renewable energy through capacity markets: An analysis of the Russian support scheme", *Energy Policy*, 46, 68-77.

[2] Ciaccia G., Doni N., Fontini F., 2010, "Auctioning wind power sites when environmental quality matters", *Energy Policy*, 38/4, 1734-1740.

[3] Darghouth, N. R., Barbose, G., & Wiser, R., 2011), "The impact of rate design and net metering on the bill savings from distributed PV for residential customers in California", *Energy Policy*, 39, 5243-5253.

[4] Ferrey, S., 2004, "Net zero: Distributed generation and FERC's midAmerican decision", *The Electricity Journal*, 10, 33-42.

[5] Fokaides P.A., Angeliki Kylili A., 2014, "Towards grid parity in insular energy systems: The case of photovoltaics (PV) in Cyprus", *Energy Policy*, 65, 223-228.

[6] Fouquet D., Johansson T.B, 2008, "European renewable energy policy at crossroads - Focus on electricity support mechanisms", *Energy Policy*, 36/11, 4079-4092.

[7] Kalkuhl M., Edenhofer O., Lessmann K., 2013, "Renewable energy subsidies: Second-best policy or fatal aberration for mitigation?" *Resource and Energy Economics*, 35/ 3 217-234.

[8] Klessmann C., 2009, "The evolution of flexibility mechanisms for achieving European renewable energy targets 2020 - ex-ante evaluation of the principle mechanisms", *Energy Policy*, 37/11, 4966-4979.

[9] Klessmann C., Lamers P., Ragwitz M., Resch G., 2010, "Design options for cooperation mechanisms under the new European renewable energy directive", *Energy Policy*, 38/8 4679-4691.

[10] Langniß O., Diekmann J., Lehr U., 2009, "Advanced mechanisms for the promotion of renewable energy - Models for the future evolution of the German Renewable Energy Act", *Energy Policy*, 37/4, 1289-1297.

[11] Lim XL., Lam WH., 2014, "Review on Clean Development Mechanism (CDM) implementation in Malaysia", *Renewable and Sustainable Energy Reviews*, 29, 276-285.

[12] Lucas H., Ferroukhi R., Hawila D., 2013, "Renewable Energy Auctions in Developing Countries", *International Renewable Energy Agency* (IRENA).

[13] Network for New Energy Choices, 2009, "Freeing the grid - Best and worst practices in state net metering policies and interconnection procedures", New York: www.newenergychoices.org.

[14] Parmesano, H., 2003 "Standby service to distributed generation projects: The wrong tool for subsidies", *The Electricity Journal*, 10, 85-92.

[15] Pollitt M.G., Haney A.B., 2013, "Dismantling a Competitive Electricity Sector: The U.K.'s Electricity Market Reform", *The Electricity Journal*, 26/10, 8-15.

[16] Poullikkas A., 2009, "A decouple optimization method for power technology selection in competitive markets", *Energy Sources, Part B*, 4, 199-211.

[17] Poullikkas A., 2009, "Economic analysis of power generation from parabolic trough solar thermal plants for the Mediterranean region – A case study for the island of Cyprus", *Renewable and Sustainable Energy Reviews*, 13, 2474-2484.

[18] Poullikkas A., 2013, "Optimization analysis for pumped energy storage systems in small isolated power systems", *Journal of Power Technologies*, 93, 78-89.
[19] Poullikkas A., Kourtis G., Hadjipaschalis I., 2011, "A hybrid model for the optimum integration of renewable technologies in power generation systems", *Energy Policy*, 39, 926-935.
[20] Poullikkas, A., 2009, "Parametric cost-benefit analysis for the installation of photovoltaic parks in the island of Cyprus", *Energy Policy*, 37, 3673-3680.
[21] Poullikkas, A., Kourtis, G., & Hadjipaschalis, I., 2012, "An overview of the EU Member States support schemes for the promotion of renewable energy sources", *International Journal of Energy and Environment*, 4, 553-566.
[22] Rokach, J., 2011, "Net metering needs safety net", *The Electricity Journal*, 11, 106-108.
[23] Sonneborn C.L., 2004, "Renewable energy and market-based approaches to greenhouse gas reduction - opportunity or obstacle?", *Energy Policy*, 32/16, 1799-1805.
[24] Varnavas C., Poullikkas A., 2013, "A Measure of Capacity Contribution of Static Mono-Si Photovoltaic Systems", *Conference Papers in Energy*, Article ID 526320.
[25] *Wien Automatic System Planning (WASP) Package: A Computer Code for Power Generating System Expansion Planning Version WASP-IV with User Interface User's Manual*, 2006, International Atomic Energy Agency, Vienna.
[26] Yin, H., Powers, N., 2010, "Do state renewable portofolio standards promote in-state renewable generation?" *Energy Policy*, 38, 1140-1149.

INDEX

D

E

F

G

H

I

N

O

P

Q

R

S

T

U

V

W

X

Y